Positive Linking

Paul Ormerod is the author of *The Death of Economics*, *Butterfly Economics* and *Why Most Things Fail*. He studied economics at Cambridge and his career has spanned the academic and practical business worlds, including working at the *Economist* newspaper group and as a director of the Henley Centre for Forecasting. He is a Fellow of the British Academy of Social Science and has been awarded a DSc *honoris causa* for his contribution to economics by the University of Durham.

Further praise for Paul Ormerod:

'It is difficult to find a reflective economist who is happy with matters as they are. Professor Paul Ormerod has gone one step further . . .' Will Hutton, *Guardian*

'Mr Ormerod not only writes about capital, but also points to a way it may at last be understood.' *New York Times*

'Ormerod is on to something . . . The tools of economics are too useful to be squandered on shadow plays.' *Business Week*

'Ormerod wants economists to stop thinking about how the world ought to behave and start looking at how it does.' Robin Blake, *Independent*

'Interesting and entertaining . . . The scale and breadth of Ormerod's analysis deserves commendation.' Adrian Woolfson, *Nature*

Positive Linking

How Networks Can Revolutionise the World

Paul Ormerod

faber and faber

First published in 2012
by Faber and Faber Limited
Bloomsbury House
74–77 Great Russell Street
London WC1B 3DA

Typeset by Faber and Faber Limited
Printed in the UK by CPI Group (UK) Ltd, Croydon, CR0 4YY

A CIP record for this book
is available from the British Library

ISBN 978–0–571–27920–3

2 4 6 8 10 9 7 5 3 1

Contents

Illustrations

Preface

My previous book, *Why Most Things Fail*, was published in the mid-2000s. I addressed what is probably *the* most fundamental feature of both biological and human social and economic systems. Species fail and become extinct, brands fail, companies fail, public policies fail. Despite the rather gloomy title, the book did well. It was a *Business Week* US Business Book of the Year.

The book built on general themes which I had already explored in my two previous books. In *The Death of Economics* in the mid-1990s I argued that conventional economics views the economy and society as machines, whose behaviour, no matter how complicated, is ultimately predictable and controllable. The financial crisis of the late 2000s showed only too clearly how deeply flawed is this view of the world, embraced enthusiastically by mainstream economists, international bodies such as the International Monetary Fund, central banks and politicians the world over. On the contrary, the economy is much more like a living organism.

In *Butterfly Economics* in the late 1990s, I developed this theme. I analysed a wide and seemingly disparate range of economic and social questions, seeing them as analogous to living creatures whose behaviour can be understood only by looking at the complex interactions of their individual parts.

Why Most Things Fail drew inspiration from the biological sciences even more, demonstrating close parallels between, for example, the extinction of biological species in the fossil record and the extinction of companies. The idea that economics should look to biology for intellectual inspiration is a long and distinguished one. Alfred Marshall, who founded the faculty of economics at Cambridge University around 1900, was the first major scholar to articulate this view.

Vernon Smith, in his economics Nobel Prize lecture in 2003, stated bluntly: 'I urge students to read narrowly within economics, but widely in science. Within economics there is essentially only one model to be adapted to every application . . . The economic literature is not the best place to find new inspiration beyond these traditional technical methods of modelling.' I have followed this precept in this book. In addition to biology, I draw on powerful insights from, amongst other disciplines, psychology and anthropology.

Both *Butterfly Economics* and *Why Most Things Fail* were fundamentally based on the concept of networks, the idea that individuals do not operate, as conventional economics assumes, in isolation, but are connected together in society. Both the theory and practice of networks is a rapidly developing area, and I make use of results at the forefront of knowledge in this book.

Viewed from a network perspective, many aspects of our social and economic world look completely different than they do from the conventional view of mainstream economics. The world is not by any means a machine whose behaviour is predictable and controllable by pulling a lever here, by pressing a button there. The individual components – people, firms, regulators, governments – interact with each other and each component has the capacity to change directly how other components behave.

In many ways, this makes successful policy making, whether in the public or private sectors, much harder. So much is contingent on who influences whom on a network and when. Simple causal relationships between a change in policy and any given outcome no longer exist – if they ever did!

At the same time, far more effective policy making becomes possible. It requires a fundamental change of mindset by policy makers. This does not mean no government, but it certainly means much more thoughtful government instead of the complacent, tick box mentality which currently dominates the public sector in the West. The positive aspects of the huge recent increase in knowledge about social and economic networks open up new possibilities for solving many long-standing problems. A lighter, smarter touch, one which exploits the positive linking aspects of our modern, networked world.

I am grateful to a large number of people for encouragement and discussions which have helped to develop the ideas of this book, and in particular to Alex Bentley, Greg Fisher and Bridget Rosewell. Julian Loose of Faber and Faber has once again proved to be a very helpful and inspirational editor.

Paul Ormerod
London, Wiltshire and Red Lumb, October 2011

Introduction

Modern economic theory was first set out on a formal basis in the late nineteenth century. There have certainly been developments since then, but at heart the basic view in economics of how the world operates remains the same. Economics is essentially a theory of how decisions are made by individuals, of what information is gathered and how it is used by the decision maker.

All scientific theories, even quantum physics, are approximations to reality. Theories involve making assumptions, simplifications, to enable us to understand problems better. A key feature of a good theory is that its assumptions are a reasonable description of the real world.

In the early twenty-first century, just as it did in the late nineteenth, economics in general makes the assumption that individuals operate autonomously, isolated from the direct influences of others. A person has a fixed set of tastes and preferences. When choosing amongst a set of alternatives, he or she compares the attributes of these alternatives and selects the one which most closely corresponds to his or her preferences.

At first sight, this may seem quite reasonable, indeed even 'rational', as economists choose to describe this theory of behaviour. But there is a serious problem with the assumption that individuals operate in isolation from each other, that their preferences are not affected directly by the decisions of others. The social and economic worlds of the twenty-first century are simply not like this at all. We are far more aware than ever before of the choices, decisions, behaviours and opinions of other people. In 1900, not much more than 10 per cent of the world's population lived in cities. Now, for the first time in human history, more than half of us live in cities, in close, everyday proximity to large numbers of other people. In the last decade or so, the internet has revolutionised communications in a manner not experienced since the invention of the printing press in the mid-fifteenth century.

The assumption that people make choices in isolation, that they do not adopt different tastes or opinions simply because other people have them, is no longer sustainable. Perhaps – perhaps, and it is a big 'perhaps' – over a hundred years ago this might not have been a bad assumption to make. But no longer.

The choices people make, their attitudes, their opinions, are influenced directly by other people. The medium via which this influence spreads is the social network. Often, social networks are thought of as purely a web-based phenomenon: sites such as Facebook. These can indeed influence behaviour. But it is real-life social networks – family, friends, colleagues – that are even more important in helping us shape our preferences and beliefs, what we like and what we do not like.

Network effects, the fact that a person can and often does decide to change his or her preferences simply on the basis of what others do, pervade the modern world. Throughout history, a crucial feature of human behaviour has been our propensity to copy or imitate the behaviours, choices, opinions of others. We can see it in the fashions in pottery in the Middle Eastern Hittite Empire of three and a half millennia ago. And we can see it today in the behaviour of traders on financial markets, where the propensity to follow the herd can lead all too easily to the booms and crashes we have lately experienced. Scientists such as Robin

Dunbar have argued that our anomalously large brain (compared to other mammals) evolved precisely because, from an evolutionary perspective, copying is a very successful strategy to follow.

This concept is just as crucial for companies and markets as it is for people. In September 2008 Lehman Brothers went bankrupt, precipitating a crisis which almost led to a total collapse of the world economy and a repeat of the Great Depression of the 1930s. It was precisely because Lehman was connected via a network to other banks that made the situation so serious. Lehman's failure could easily have led to a cascade of bankruptcies across the world financial network, first in those institutions to which Lehman owed money, then spreading wider and wider from these across the entire network. Incredibly, neither the systems of financial regulations which were in place, nor the thinking of mainstream economics which influenced policy so strongly, took any account of the possibility of such a network effect.

A world in which network effects are a driving force of behaviour is completely different from the world of conventional economics, in which isolated individuals carefully weigh up the costs and benefits of any particular course of action. A world in which network effects are important is a much more realistic description of the human social and economic realities which exist in the twenty-first century. It is the implications of this world which I explore in this book.

Incentives have not disappeared as a driver of human behaviour. It is still the case that if, say, Pepsi raises its price compared to Coke, more Coke and less Pepsi will be sold. This is the world which economic theory describes. It is not wrong. But it is often misleading, for it offers only a very partial account of how decisions are made in reality. Network effects can be far more powerful than incentives, and we will see many examples in which network effects have completely swamped the impact of incentives, leading to outcomes completely different from those intended by policy makers.

Network effects require policy makers, whether in the public or corporate spheres, to change radically their view of how the world operates. In part, they make policy much harder to implement successfully, and they help explain many of the failures of policies based on the assumption that incentives and not network effects are the key drivers of behaviour. But they open up the possibility of much more effective and successful policies, ones which harness our knowledge of network effects and how they work in practice. Hence the main title of this book: Positive Linking.

1

Unintended Consequences

On Wednesday 16 October 1555, two of the leading members of the reformed English Protestant Church, Hugh Latimer and Nicholas Ridley, were chained to a stake in the city of Oxford. They were then burned to death. By what amounted to a series of historical accidents a Catholic, Mary, had become Queen, the ruler of all England, scarcely two decades after the Church of England made its historic break with the papacy. She was attempting to re-impose Catholicism by a policy of publicly burning leading Protestants. If they renounced their faith, their lives would be spared and they might even continue to enjoy the power and trappings of high office. If not, they faced the fire.

But far from quailing at a terrible fate, Latimer and Ridley embraced it cheerfully. 'Be of good comfort, Master Ridley, and play the man; we shall this day light such a candle, by God's grace, in England, as I trust shall never be put out,' Latimer allegedly pronounced. They, along with other condemned Protestants, had formed a deliberate policy of facing death with equanimity, in order to make a positive impression on those who witnessed the burnings.

They believed that the story of their end would spread by word of mouth far beyond those present at the executions. Existing Protestants might be encouraged by their example to be steadfast in their faith, and new converts gained. And on this occasion, the martyrs were ultimately proved to be correct. On Mary's death, Protestantism was restored as England's religion.

Flash forward over 400 years to another event in English history, far less momentous, but one which offers a vignette of popular culture, not of the mid-sixteenth century but the late twentieth.

In Sardinia during the 1990 soccer World Cup, the English supporters were feared for their violent reputation. One evening in Cagliari, a large number gathered in the streets. Facing them were the police. As Bill Buford relates in his excellent book *Among the Thugs*, various individuals made attempts to stir the fans into collective action without success. Making himself conspicuous so that others could see his actions, one threw a metal object at the police. Another charged the police and yelled for others to follow. Further attempts were made by isolated fans to encourage the crowd into collective action, but none joined in.

Tiring of the whole situation, and in response to the actions of one particular youth, a police captain fired his pistol into the air in a signal for the potential mob to disperse. The reaction was unexpected. At the sound of live ammunition being discharged, the English supporters immediately began to destroy property and attack the police. The very action intended to subdue the fans into quiescence provoked exactly the opposite reaction. The individual supporters suddenly turned into a mob.

These two stories, disparate though they may seem, have a great deal in common. They illustrate the seemingly perverse and apparently irrational ways in which people can behave. Rather than sullenly dispersing back to the safety of their hotels or into bars when a firearm was discharged, the soccer fans ran at the police. Latimer and Ridley were offered the choice not just of their lives but their freedom if they embraced Catholicism. Rather than meekly agreeing, whatever private reservations might have remained, they chose to suffer an appalling death. Not just history but contemporary life is replete with examples of people behaving in seemingly inexplicable ways.

A key theme of this book is that these widespread forms of behaviour are explicable. They are illustrations of the power of social networks. Today, the phrase 'social networks' is often synonymous with networking across the web on sites such as Facebook. But this is just one, albeit new and important, aspect of a phenomenon that has existed for centuries. People do not live in isolation, but in society. Their lives are filled with interactions across social networks. The network of their families, the network of their work colleagues, the networks of their hobbies. Real-life social networks in which people meet, gossip, chat, argue. Networks in which people's choices, behaviour, opinions can be influenced, shaped, even altered dramatically by the process of social exchange with other people.

Within these social networks, people often copy or imitate what others do or think, for a variety of motives. An individual might have formed a private view on a matter, but might believe that others with a different opinion are better informed and so changes his or her mind as a result. Or someone may accept the behaviour of a particular social group simply from a desire to conform. More subtly, peer acceptance might give an individual permission to behave in a way that, in a different social context, would be unacceptable.

The idea that copying is an important aspect of behaviour does not mean that individuals operate as automatons, that they have surrendered control over their decision to others. People can copy and still retain a clear sense of agency, of purpose and intent over their own actions. So in a strange city, you may consciously decide to copy others, to go to the restaurant where there are lots of customers rather than to the one in the same street where there are few. Lacking any other reliable information, lacking local knowledge yourself, you decide to be influenced by the choices made by others. Even in the highly connected world of the twenty-first century, networks are not everything. People still retain their individuality, their capacity to decide actions and beliefs for themselves, despite what is popular, either in society as a whole or amongst their particular group of friends, family or work colleagues.

Most public policy on social and economic matters is based on the premise that people, or indeed companies, behave as individuals when they are making decisions. Like so many Robinson Crusoes, people exist in splendid isolation. And it is this view of the world that is epitomised by mainstream economic theory.

We explore in this book the connection between the impact of incentives, of the assessment of costs and benefits of different actions, on individuals, and the effect of social interaction across networks. When the power of the network takes over, people are no longer acting autonomously, but as part of a social group, and their behaviour and decisions are driven by the process of copying, of imitation.

Sometimes the initial impact of changes to incentives on the behaviour and decisions of a few individuals will be seen to be enhanced as the power of the social network takes over, and this effect can on occasion be dramatic. But, equally, there are times when the impact of copying behaviour across a social network, of imitating the behaviour of others, does not just offset the effect of incentives, but takes the system in the completely opposite direction to what was intended.

In recent decades, the discipline of economics has exhibited powerful tendencies of intellectual imperialism. Not content merely to analyse the familiar areas of firms, consumers, prices and markets, economists have turned their attention to a wide range of social issues, seemingly far away from the original scope of economics: the study of the allocation of scarce resources. The institution of marriage, crime, piracy, drug addiction – economists now focus on all of these and more.

8

Indeed, the two historical vignettes which opened this chapter can be translated into the context of economic theory. The popular image of economics is that it deals with 'big' things, national output (GDP), unemployment, inflation, interest rates. This is macroeconomics and these are the topics which appear in the newspapers and on our television screens and on which economists are regularly seen to pronounce.

But, in essence, economics is a theory about how individuals make decisions. About decisions made at the microeconomic level. The measure of the relative importance of microeconomics is indicated by the fact that, over the past two decades, the Nobel Prize in economics has been awarded for work which has been unequivocally 'macro' in character on only four occasions. Not all the others have gone to micro, for some of the Laureates have made advances in techniques of statistical analysis, but micro distinctly outweighs macro in these awards. At the core of microeconomics is a series of theoretical postulates about how the so-called 'rational' individual makes decisions. For example he or she has a well-defined and fixed set of preferences concerning the choices on offer. He or she gathers all available information when making a decision, matches it against his or her preferences, and then makes the best possible decision – the 'optimal' decision, as economists like to say to give it a more scientific air – given the information and the preferences.*

So the agent – the jargon phrase in economics for the person making the decision – may, if the products have the same price, prefer Pepsi to Coke. (It is in fact rather useful to use 'agent', rather than 'he or she', since the word subsumes the two genders and avoids having to repeat the two.) But if the price of Pepsi rises relative to Coke, at some point any given agent will

---

* I have discussed this model in more detail in previous books, such as *The Death of Economics*, a critique of free-market economic theory written in the mid-1990s.

switch and buy Coke instead. This is not because the agent's preferences have fundamentally altered, but because the money saved in buying Coke rather than Pepsi in this illustrative example can be used to buy more of other products. So, overall, the preferences of the agent might be more closely matched by switching to Coke.

Another example is the British savoury spread Marmite. The only other countries where I believe either it or a close variant are on general sale are Britain's closest cultural neighbours, Australia and New Zealand. Based as it is on the scrapings of the fermented residue at the bottom of beer barrels, agents' preferences on this tend to be sharply divided. I cannot abide it. My wife adores it. But there is some price at which I could be persuaded to eat it, probably a negative one in which the producer paid me rather than the other way round.

When making choices between fairly straightforward, inexpensive, well-established consumer products, the economist's view of 'rational' choice may be reasonable. Coke, Pepsi, Marmite have all been around for a long time and agents have formed their preferences. They are unlikely to suddenly alter them, at least in any appreciable numbers.

In making a decision about the choices on offer, only a relatively small amount of information needs to be gathered, mainly concerning price. This latter factor might not be completely obvious, because the price per unit of weight or volume may vary from store to store, by pack size, or because of special offers such as 'buy one, get one free' or 'three for the price of two'. The mathematical capabilities of many people are known to be low. For example on the very day I write these words, a TV advert to recruit teachers created by the Training and Development Agency for Schools has been exposed by a fifteen-year-old schoolboy as containing the wrong answer to a fairly straightforward question. But even making allowances for this, there is a limited amount of information for agents to gather before matching it to their preferences.

This model of rational behaviour is no longer relevant in many circumstances in the world of the twenty-first century. Agents face a vast proliferation of choice, massive information overload. Many of the products on offer are highly sophisticated, difficult to evaluate in terms of their attributes. And we live in a world which is far more connected, in which we are far more aware of the opinions and behaviour of others, than we were a hundred years ago when standard economic theory was first being formalised. In 1900, the clear majority of the population of the world lived in relatively isolated villages. In the twenty-first century, the majority lives in cities, in close proximity to large numbers of other people. And the revolution in communication technology brought about by the internet makes us dramatically more aware of the behaviour of others than at any time in the whole of human history.

We need a new model of rational behaviour, one which is empirically consistent with the real world, the world of the twenty-first century. The economist's definition of rational behaviour is only one possible way to define the concept of rationality. Behaviour which does not follow the precepts of economic rationality is *not* irrational, as economists would have us believe. Indeed, in the modern world in many contexts it is the economic definition of rationality which has become irrational!

The development of such a view of the world, a more realistic view of how agents actually behave in the social and economic contexts of the twenty-first century, is a main theme of the book. It has radical implications for the conduct of policy, both corporate and public. Potentially, its impact is very positive. Our knowledge of how networks influence behaviour in the social and economic worlds is growing rapidly, both theoretically and empirically. The opportunity both to exploit this knowledge and to develop it even further, for there is much still to be done, over the coming decades is enormous. Successful policy making in the highly connected, networked world of the 21st century will be impossible without understanding positive linking, how we can use what are often abstract and difficult concepts to help shape a better world.

*

How, then, might we think about the two historical episodes described above in the context of standard economic theory? The theory makes claim to be a general description of human behaviour, a general theory of how people make choices. These examples may seem outside the conventional areas of economics, but if a theory is claimed to be general, it ought to be able to illuminate these events. Besides, it is economists themselves who have pushed the theory into areas such as marriage and crime and claimed that it has strong explanatory power in what might more usually be thought of as social rather than economic settings. The Chicago economist Gary Becker received the Nobel Prize for exactly this kind of work.

The individual preferences of the soccer fans were to have some sort of riot in which property would be vandalised and innocent passers-by made to cower in fear or, even better, injured in some way or other. This is why they had assembled as they did. A message had been passed to meet at a particular time, six o'clock as it happens, in a particular square. The colloquial phrase used was 'it's going to go off', meaning that, for those interested, the gathering would offer an opportunity to participate in creating mayhem in the city of Cagliari.

The fans would derive 'utility', again using the jargon of economics, from rioting. But what were the related costs to set against these 'benefits', using the phrase in inverted commas on this occasion to emphasise that these were, of course, benefits only to those involved in trashing the city, not to those unfortunates on the receiving end. The most obvious cost was the phalanx of police standing in front of them. Heavily equipped with helmets, shields, truncheons and guns, they were clearly capable of inflicting costs, such as a beating or arrest and prison, on anyone foolish enough to provoke them.

Economic theory usually allows individuals to differ in their preferences. Incredibly, as we shall see later in the book, the trends in macro theory in recent decades have been to suppress this, trying to explain the economy as a whole in terms of a single 'representative agent'. But more of this later. For now, we remain firmly in the terrain of microeconomics, where agents can have different preferences. The youths gathered in the square would undoubtedly differ in the benefit each individual believed he would gain from having a riot compared to carrying out other activities, such as having a beer or reading a book on Einstein's theory of general relativity or Shakespearean sonnets. They would differ in the evaluation of the costs of any police action inflicted on them. And, according to standard economic theory, they would even be allowed to differ in their assessment of the probability of being the recipient of such action themselves (with the strict proviso, and I am not making this up, that over the course of a series of such riot events, each fan on average assesses the probability correctly).

On this view of the world, every single one of the fans who responded to the verbal message 'it's going to go off' derived utility from participating in hooliganism. On arriving in the square and seeing the police, at some point they must have formed the view that these benefits outweighed the likely costs, or the riot would not have happened. It is possible that a few crept discreetly away, having come to the opposite view, but Bill Buford's description of the events certainly suggests that almost all the English fans present participated in the subsequent vandalism and general criminal behaviour which occurred.

But this does not take us very far in understanding why the riot started when it did. The fans stand, confronting the police on a hot late afternoon. But at first, they are a collection of individuals and not a collective mob. They have the potential to become a mob, but nothing happens. Several fans try to incentivise them all to start behaving badly by carrying out prominent acts of bravery, or lunacy as most people would see it, against the police. We can in fact readily understand the behaviour of these particular individuals from the point of view of economic theory. Considerable status would be attached to being seen as a leader by the other fans, being seen as a Top Boy, to use the British colloquial expression for the leader of a gang of thugs or hooligans. The benefit from this would be perceived as outweighing the undoubted increase in the potential cost to the individual by identifying himself so prominently to the forces of law and order.

This whole rationale for the event, as described by conventional economic theory, may already seem somewhat convoluted, but it now becomes even more so. Why did the fans as a whole not respond to the actions of the individuals who deliberately tried to incite a riot by their provocative actions? According to this theory of how agents behave, we have to suppose that the responses to these by the police were such as to temporarily tip the balance between costs and benefits in the minds of all the would-be rioters. In other words, when a youth came forward and threw a metal object at the police, they perhaps brandished their truncheons more fiercely, and this signal increased the likely costs of a riot in the minds of the fans.

But then the police captain, tiring of confronting this unpleasant group of badly dressed, smelly individuals,* fired live ammunition into the air. He clearly believed that signalling this potential cost – the possibility they would be fired upon – would be a sufficiently large incentive to make them eschew the pleasures of a riot on this occasion. The cost of being the recipient of a bullet surely outweighs that of even a savage truncheoning, but the response of the fans suggests otherwise.

The fans immediately charged the police. A possible reconciliation with the core model of individual behaviour in economic theory is that the shot fired into the air was a sign of weakness on the part of the police, a sign that they would not actually open fire on the English, regardless of what they did. But this argument is now getting pretty tenuous. Ex post, economic theory can rationalise almost anything which has ever happened, but these attempts often amount to no more than a Just So story, as is certainly the case here. Their credibility gets stretched well beyond breaking point.

A much simpler explanation can be given in terms of networks. When the fans first gathered in the square, it is not implausible to interpret their behaviour in terms of individualistic economic theory. Each of them enjoyed a riot, they gained 'utility' from it, in the jargon of economics. As noted above, the strength of their individual preferences for participating in a riot compared to other activities undoubtedly varied, as did their assessments of the costs. The delay between the fans assembling and the riot starting, and the lack of collective response to the efforts of reckless individuals to incite them, suggest that almost all of the supporters had formed the view, given the serried ranks of the forces of law and order confronting them, that the potential costs involved outweighed the benefits.

* *Among the Thugs* graphically illustrates these qualities in a variety of contexts.

A shot was fired. The collection of individuals immediately became a mob. They lost their individual identities. And their preferences altered dramatically, so that when the gun went off they charged the police, acting as a single unit. They had arrived as individuals as part of a social network with a shared interest in hooliganism. Information about a potential outlet for this activity had been passed across this network. But the action of the unfortunate police captain altered qualitatively the structure of this network. The individuals became fused as one, with an overwhelming preference to riot almost regardless of costs to themselves as individuals.

What scientists call a 'phase transition' had taken place. When water is gradually cooled, it remains water as the temperature drops from ten to nine to eight degrees and so on. Then, suddenly, as it passes through zero, a phase transition occurs. Water becomes ice.

A simple example of this phenomenon which is almost certainly more familiar to most readers of this book than taking part in a public riot is a social gathering with friends, in a bar or perhaps at a party. Each individual present enjoys alcohol in moderation and dislikes hangovers. But the company is delightful, the wine flows. The collective mood temporarily overcomes the preferences of individuals. And the effect of the social network present in this particular milieu is that almost everyone is induced to drink more than he or she intended at the start of the evening, or would drink on their own or in a smaller or less congenial group. The next day, operating once more as individuals with their individual preferences restored, some will undoubtedly regret the collective set of preferences – to consume yet more alcohol – which spread across the social network. One or two may go so far as to pledge to themselves never to drink again. (Until the next time, of course!)

James Surowiecki wrote a very interesting book in 2004, *The Wisdom of Crowds*. This is essentially about the process of answering a question by taking into account the opinions of a large number of individuals rather than relying on just a few, no matter how expert these people might be. There are many practical examples where the 'crowd' certainly gives a more accurate estimate, such as the classic ones of guessing the number of sweets in a jar or the weight of a prize bull at a country fair. The word 'crowd' is put in inverted commas here, because the process only really works when the individuals participating in the process remain as individuals and not part of a crowd in the way the soccer vandals were.

The crucial assumption needed for the average of the collection of individual opinions to be more accurate than a single expert is that they do indeed form their views independently, without reference to those of others. Once this independence vanishes, once the agents become fused into a single whole, often the outcome is not so much the wisdom, more the 'madness of crowds', as it was described by Charles Mackay as long ago as 1841.

*

There is another very general point, and another key theme of this book, to take from the example of the English soccer fans and their rampage through the streets of Cagliari. The individuals received several attempts to incite them. But none of these worked. Then, completely unexpectedly, the one event which any detached observer might think would offer a clear deterrent, the firing of the gun, turned the group of individuals into an enraged crowd.

And this is the point. Most attempts to spread a choice, an opinion, a type of behaviour, across a network of individuals fail. The events in Sardinia are simply an example of this general point. And it is why we have to be very careful when designing public policy. Like the police captain, policy makers will often have little idea about the likely consequences of their attempts at nudging groups towards particular decisions or opinions. Duncan Watts, formerly Professor of Mathematical Sociology at Columbia, now director of the Human Social Dynamics group at Yahoo! and someone we will meet in much more detail later, used a phrase for this fundamental property of networks. They are 'robust yet fragile'.

The collection of individuals who make up a network will, most of the time, exhibit stability with respect to most of the 'shocks' which this particular system receives. The shock could be a piece of news in the context of financial markets, an advertising campaign in a consumer market, or, as here, attempts to incite a group of fans to alter their preferences and attack the police. The system is stable in the sense that most shocks make very little difference, they are absorbed, shrugged off, and few people change either their minds or their behaviour as a result. So the network is 'robust'.

But, every so often, a particular shock may have a dramatic effect. So the network is also 'fragile'. The behaviour of individuals across the whole, or almost the whole, of any particular network might be altered. Before the event, it can be very difficult if not impossible to discover what the eventual impact is going to be. A big shock, almost by definition, will have big consequences. So if the Italian police had opened up with machine-gun fire directly at the fans, we could reasonably conclude that in this particular instance a riot would not have taken place. But most events, most attempts to change behaviour, do not fall into this category. Most have very little impact. But occasionally, one does.

From these sordid events in Sardinia, we can now return to the altogether more dramatic happenings in Oxford over 450 years ago. Again, we can offer a partial explanation in terms of incentives, of costs and benefits, when people are acting as if they were isolated individuals. But, again, this kind of rationale soon becomes incomplete. Networks are needed to complete the picture.

Since time immemorial religion has been, and continues to be outside Western Europe, a major presence in human society. Yet mainstream economics has virtually ignored the topic, certainly in comparison to the enormous amount of work carried out in sociology, anthropology, psychology, history – disciplines considered 'soft' by most economists.

There has been some work on religion in economics. But when Laurence Iannaccone of George Mason University in Fairfax, Virginia, probably the leading modern economic scholar in this area, wrote an 'Introduction to the Economics of Religion' in the prestigious *Journal of Economic Literature* in 1998, he noted that 'the study of religion does not yet warrant a *JEL* classification number'. This simple observation is significant in revealing the amount of attention paid to religion by economists up to the late 1990s.

The example of the Oxford Martyrs is specifically religious, but the arguments being considered are relevant much more generally to all human belief systems where faith or ideology is important. In the decades around the middle of the twentieth century, why did highly placed individuals in both America and Britain decide to give their loyalties to the ideology of communism and betray their countries by revealing secrets to the Soviet Union? Neither Kim Philby nor Alger Hiss, two of the most notorious spies, appears to have been motivated by money, by the set of standard incentives in the economist's toolkit. They were motivated by faith, the wholly misplaced faith that the Soviets would create a better future for all humanity. They were utterly and completely wrong. But they believed.

History is replete with examples of ideological differences which cannot be accounted for on the basis of 'rational' economic decision making. Given the historical importance of religion, many such disputes involve this topic. But thinking still of the Soviet Union, after the collapse of tsarist rule in 1917, a vast proliferation of competing political ideologies bubbled to the surface. The Western liberalism of Kerensky, who formed a government for a few brief months in 1917. The several varieties of Whites, believers in monarchy, against whom the Bolsheviks fought a brutal and debilitating civil war. Within the revolutionaries themselves were differing ideological tendencies: anarchists, social revolutionaries, Mensheviks, Bolsheviks, to name but a few. And Bolshevism, the ultimately victorious faction embodied in the Communist Party of the Soviet Union, was notorious for vicious internal ideological disputes even when the key authority figure of Lenin was still alive.

There were undoubtedly many motives at work in each of the arguments and struggles. Personal ambition mixed with genuine belief that your opinions and those of your faction were the correct ones, the ones which would bring about Paradise on Earth. But all of them involved faith and ideology rather than rational, incentive-based decisions. So, although we now resume the discussion focused on religion, we should keep in mind that the points are relevant to any faith- or ideology-based dispute in human affairs.

*

All sciences classify the various aspects of their discipline. We earlier came across the basic distinction between micro- and macroeconomics. But the scientific classification goes into much finer detail than this. The *JEL* (*Journal of Economic Literature*) system is the one used by all economists. It divides the subject into well over 500 sub-categories. And in the late 1990s religion, one of the most fundamental features of human society, did not warrant a category of its own, so little work had been done on it. The situation has now changed. Religion does have its own economics sub-category. But, revealingly, it is allocated in section Z, 'Z12' no less, coming right at the end of the very long list, lower down even than the ten sub-categories in category Y, all of them 'Miscellaneous Categories'.

In some ways this is surprising, given that Adam Smith, the founding father of modern economics, wrote about the topic extensively in one of his two great books, the *Theory of Moral Sentiment*. He even analysed religious issues from the perspective of agents responding to incentives in his *Wealth of Nations*. He discussed how the clergy could be motivated by self-interest, how monopoly is as bad for religion as it is in other areas of human activity, arguing that competition – being able to choose between competing religions – is good.

Economics had to wait almost exactly 200 years before Smith's analysis was extended, in a model developed by Corry Azzi and Ronald Ehrenberg and published in the top-ranking *Journal of Political Economy*, based in the free market-oriented University of Chicago. A short summary gives a flavour of both this particular model and subsequent work by economists on the topic of religion.

Individuals allocate their time and money amongst religious and secular commodities with the aim of maximising lifetime *and* afterlife utility. 'Afterlife consumption', as Azzi and Ehrenberg describe it, is the primary goal of religious participation. Secular utility depends in the standard way on inputs of time (work) and the products which are purchased. Afterlife utility depends upon the entire effort devoted by the individual to religious activities over his or her lifetime.

The article is not a spoof, though it would be quite difficult to invent a more effective satire of the model of utility-maximising Rational Economic Man which dominates the entire literature of economics. So many points spring to mind. For example, on a purely technical point within the spirit of the literature itself, but one which is important empirically in some main religions, 'afterlife utility' does not vary continuously with the amount of effort devoted to religion during your lifetime. It is a simple binary outcome: either you are in Heaven, with boundless pleasure, or in Hell, with endless pain. And the outcome might very well not depend on the amount of time and effort which you devote to religious activity. Who will be saved in the well-known parable, the self-righteous Pharisee, obsessed not only with his own virtue but with the constant public display of it, or the sinful but repentant publican who devotes very little time and effort to religion? To be fair to the economics of religion, it has moved on to consider participation more as a group activity and to focus on institutions and their behaviour. But it has very little to say about the most fundamental question: why believe at all?

Despite all this, incentives were certainly at work in the religious world of England in the 1550s. Although there were many nuances within each religion, individuals faced a basic choice between being Catholic or Protestant. Queen Mary's father, Henry VIII, had broken with the Pope in the 1530s and established the Church of England. The institution had gradually come under the control of hard-line Protestants, a trend which accelerated during the short reign of his young son, Edward VI, in the years around 1550. Following Edward's premature death from tuberculosis, Mary – forever known in English iconography as 'Bloody Mary' because of her burning of the martyrs – had come to the throne determined to restore Catholicism.

There were some important directly economic issues to settle. Henry had carried out the biggest seizure of private property in English history, when the monasteries were dissolved and their lands confiscated by the Crown. Under Edward, the church leaders had gone even further, stripping and looting churches of the elaborate trappings and ornaments of Catholicism. How far should these measures be reversed? Mary's main adviser, Cardinal Pole, an Englishman who had almost become Pope in 1550, advocated a complete restoration. Mary had to balance the immediate benefits to her Church against the potentially destabilising political consequences and costs of expropriating the property which Henry had sold on almost immediately to wealthy noblemen and merchants.

But the main question facing her was how to restore the old religion of Catholicism, how to persuade people to re-embrace what she regarded as the true faith. It appears to be the case that the clear majority of the population in fact still adhered to Catholicism. Protestantism was the new brand, as it were, and was still some considerable distance from displacing the market leader. But it had achieved a strong market presence in London and its immediately surrounding areas, then as now the key focus of English political and economic power. Even more pertinently, the leaders of the Protestant Church were, in general, militants.

The bishops and other prominent churchmen could be, and were, removed from their formal positions by simple administrative acts and put in prison. A few were willing to adapt, presumably because they were attracted by the benefits of office and attached little weight to the potential afterlife costs of displaying devotion to possibly erroneous doctrines. The most notorious of these was the remarkable Anthony Kitchin, Bishop of Llandaff in Wales. He was the only person to serve as a bishop under all the various forms of religion embraced by Henry, Edward, Mary and her successor Elizabeth I and who would, in the words of one prominent historian of the time, have doubtless become a Hindu provided he could continue to remain Bishop of Llandaff. Behaviour such as this was satirised immortally in the eighteenth century popular song 'The Vicar of Bray', recounting the contortion of its eponymous subject in remaining in ecclesiastical office through the religious changes brought about by successive English monarchs. The chorus is a monument to placemen and timeservers everywhere:

> And this is Law, I will maintain,
> Until my Dying Day, Sir,
> That whatsoever King may reign,
> I will be the Vicar of Bray, Sir!

But the removal from office of most of Edward's leading clergy altered neither their beliefs nor those of lay believers in Protestantism. Mary and her advisers soon settled on a policy of terror to deal with this problem. Well-known Protestants would be given every opportunity to recant, but if they continued to refuse they would face the flames.

On the face of it, the strategy was a sensible one to follow from Mary's perspective. There had been many previous examples in human history of terror being successful in achieving its aim. And specifically in England, only 150 years previously, the Lollard heresy, an early form of Protestantism, had been suppressed effectively by a few selective burnings.

*

We might usefully pause to ask why this might be the case. The question of religious or ideological belief is enormously complicated, and one which is ultimately not susceptible to explanation by the model of 'Rational' Economic Man. In terms of this set of behavioural postulates, the agent has first of all to gather available information. But in this context, what is the relevant set of information? By definition, the existence of the afterlife can never be proved, no matter how much information we might gather. The information then has to be processed to come up with the best – sorry, the 'optimal' – choice.

The seventeenth-century French philosopher Blaise Pascal came up with his famous wager to claim that the best strategy is in fact to believe in God. Essentially, he argued that since we are incapable of knowing whether God exists or not, we have to wager on the outcome. In terms of the agent's overall happiness, the gains and losses of belief or non-belief have to be taken into account. In Pascal's own words, 'Let us weigh the gain and the loss in wagering that God is . . . If you gain, you gain all; if you lose, you lose nothing. Wager, then, without hesitation that He is.' In other words, if God exists and you believe, you gain an infinite amount of happiness but if He does not, you lose nothing. But equally if God does not exist, you lose very little either way.

Pascal was a highly original thinker, and his wager is one of the seminal contributions to modern theories of probability and decision making. Not surprisingly, there is a very large academic literature on his wager, a good introduction being in the online Stanford Encyclopedia of Philosophy.* The details need not concern us here, but suffice to say that even after thousands of academic articles, the outcome is unclear. We cannot establish an agreed basis on which an agent might use rational behaviour to believe in God, or a supreme being, or not.

Ultimately, religious or ideological belief for the individual is a matter of faith and not rational analysis. The social networks in which the person is embedded are also crucial in terms of cultural

* http://plato.stanford.edu/entries/pascal-wager/

norms and peer pressure and acceptance or otherwise of belief. 'Embedded' can mean far more than the current position. It can embrace the networks in which they grew up, for example, or which they have been part of previously and which helped shape their current beliefs. And in mid-sixteenth-century Europe, the question was: which variety of religious ideology to believe?

Even the most devout believer experiences doubt from time to time. Within Christianity, even the Apostles themselves experienced crises of faith in the immediate aftermath of the Resurrection. The two men on the road to Emmaus with Jesus were unable to recognise him, a scene given modern vibrancy in T. S. Eliot's memorable phrase in *The Waste Land*: 'who is the third who walks always beside you?' The phrase 'a doubting Thomas', meaning a sceptic, has its origin in the Apostle Thomas, who refused to believe until he had placed his hands in the Crucifixion wounds.

In any event, in the rapidly changing circumstances of sixteenth-century England, who could be really sure what was the true faith? Was the Pope the true Head of the Universal Church, or was he merely the Bishop of Rome, or even the Antichrist? The Bishop of Llandaff, whom we met above, would not have been alone in being able, if required, to subscribe to any one of these three distinct propositions.

In contrast, the prospect of being burned alive was only too real and certain. Sometimes, if the fire took hold well, death could be reasonably quick owing to oxygen deprivation, but it was an appalling end nonetheless. Equally, however, contemporary documents record examples of victims dying in prolonged agony, pleading for 'more fire' as a combination of damp wood and perverse winds slowly roasted them alive.

So Mary's policy had a definite incentive element, and agents did react. Some fled abroad. Others openly renounced

Protestantism. The most famous of all, Thomas Cranmer, Archbishop of Canterbury and head of the Anglican Church, recanted in prison no fewer than six times before finally summoning the courage to be led to the stake.

But we know from history that her policy failed. Even as Mary lay dying in the summer and autumn of 1558 after four years of terror, there was still a persistent supply of martyrs willing to be burned. And when her sister Elizabeth restored the Protestant faith, the nation embraced it with enthusiasm.

Networks were the reason for this failure. The negative incentive of the fear of the stake took her so far, but not far enough. It was overwhelmed in completely the opposite way by the power of networks. And networks were present in two separate but closely related ways.

The first was the very close network between the militant Protestants themselves, maintained even when they were in jail awaiting interrogation or execution itself. They sent messages, exchanged letters, a veritable torrent of encouragement to keep the faith and set an example to the population as a whole.

This latter, the entire people of England, constituted the second network. Much more loosely structured than the tight-knit religious one, information even in those days did pass pretty rapidly across communities. Most of the population of the country lived within three days' ride of London. The internet it was not, but news certainly travelled. And the burnings themselves were major public events, often attended by thousands of people. The Marian regime sometimes unwittingly contributed to the potential number of favourable message bearers. For the execution of Latimer and Ridley, for example, every household in Oxford was compelled to send at least one member to witness the event. The burnings were not just a deterrent. The authorities were in fact aware that they might also be a source of inspiration. So, leading

Catholic figures preached on pulpits specially constructed near the pyre, explaining the heresies of the condemned and expounding what they saw as the true faith. People could be either educated or frightened by the spectacle.

Much of our information on these events comes from a remarkable book, John Foxe's *Book of Martyrs*, first published very early in the reign of Queen Elizabeth I. The Protestant clerical elite were convinced from the outset that their deaths could cause the policy to rebound on the persecutors. They were well aware of the necessity of creating a good impression at the stake. Accounts of their behaviour would spread by word of mouth. Existing Protestants would be encouraged to keep the faith, and waverers influenced by their steadfastness.

To this end, then, they encouraged one another. On 8 February 1555, on the morning of his execution, Laurence Saunders, a noted Protestant preacher in London and the Midlands, wrote to his wife and supporters: 'God's people shall prevayle: yea our blood shal be their perdition, who do most triumphantly spill it.' He actively encouraged them to attend and enjoy the event: 'Make haste my deare brethren, to come unto me that we may be mery.'

In terms of dramatic impact, Saunders died embracing the stake. John Rogers, the first of the 300 or so martyrs to be burned, was seen to be washing his hands in the flames and Archbishop Cranmer signalled his adherence to the Protestant faith by thrusting into the fire the right hand which had previously signed his humiliating recantations. John Hooper, Bishop of Gloucester and a notoriously grumpy man, took the opportunity on his way to execution to bless a blind child and greet local dignitaries – for all the world like a modern politician out on the campaign trail for votes. Foxe put the crowd assembled to watch him burn at 7,000, 'for it was market-day and many also came to see his behaviour towards death'. Seven thousand may not seem a lot, but relative to the size of the population, it was the equivalent of around three quarters of a million in present-day America. In short, a massive crowd.

We know the outcome of this particular historical event. Incentives were put in place to persuade people to adopt one particular set of beliefs, and these had a certain amount of success. But their impact was completely offset and indeed dwarfed by the impact made by the martyrs across the network of the population as a whole. The deliberate policy of calm and even joyous acceptance of death made an impression across the land and people were influenced directly by this behaviour.

But there was no guarantee in advance that this would happen. Indeed, during the years of the Marian terror, there was evidence that the policy was working. As mentioned earlier, some leading Protestants left the country and others, from all walks and levels of life, recanted.

*

A contemporary example both of the potentially huge effect of networks and of the inherent uncertainty of outcome they create is the momentous events taking place in North Africa and the Middle East. I am writing these words in early April 2011, when neither I nor anyone else knows how they will unfold even in the (historically) short time between the writing and the publication of this book. And I am leaving the original words unchanged as the book itself is being revised during the summer and autumn of 2011. I want to capture on record how things stood in April of that year, to illustrate the uncertainties involved in such situations.

As I write, protests in some countries have already been followed by changes in regime. In Algeria, the government appears to have been able to defuse the tension. In Syria, the protests seem to be in the process of being ruthlessly suppressed. And the current situation in Libya is, to say the least, chaotic and uncertain.

The immediate catalyst to the events was Mohamed Bouazizi, a twenty-six-year-old Tunisian. Bouazizi was a university graduate living in a provincial town where he was unemployed and trying to find, but unable to get, work. He started selling fruit and vegetables in the street without a licence. The authorities put a stop to his activity, confiscated his goods and humiliated him. In response, he set himself alight and died in hospital on 4 January 2011. This sparked the riots which forced the President of Tunisia to flee the country, and was followed by similar uprisings in Egypt, Libya, Jordan, Yemen, Bahrain and Syria.

Clearly, the potential for social unrest was already in place in all these countries, ruled for decades by undemocratic regimes of various degrees of corruption and brutality, with large numbers of discontented young people. The incentives to replace the regimes were there. And doubtless, there had been many individual protests against the regimes, any one of which might have spread like wildfire across the latent network of the desire for change. But these protests have vanished into the mists of history. The network proved robust with respect to these now-unknown shocks. In the case of Bouazizi, however, the network responded in a fragile way.

Bouazizi's act of defiance was spectacular. Is this why it succeeded and others did not? Perhaps. Readers of a certain age will undoubtedly recall Jan Palach and his role in the events following the so-called 'Czech Spring' of 1968. Under the leadership of Alexander Dubček, the government of the then-Czechoslovakian state had carried out a series of liberalising measures which alarmed the Soviet leadership in the Kremlin. Czechoslovakia was part of the Warsaw Pact group of nominally independent countries, controlled in practice by the Soviet Union. In August 1968, Soviet military forces occupied the country and Dubček's government was removed from office.

A group of students made a pact to burn themselves to death in public as an act of protest against the Soviet invasion. Jan Palach actually carried it out, not in some remote provincial town but in the principal square of the Czech capital, Prague, on 16 January 1969. The event attracted widespread publicity world-wide, in contrast to the self-immolation of Bouazizi, with a leading British newspaper feeling able to write just four months after his death that 'the name of Mohamed Bouazizi has largely been lost in the unfolding story of the Arab Spring'.* Palach's self-immolation did trigger demonstrations, but these proved to be far from sufficient to bring about change and were suppressed by the security services. So in a network context, we cannot even say in advance whether or not a truly dramatic gesture such as being burned to death for your beliefs will be sufficient to persuade others and bring about change.

*

These rather disparate historical events have introduced the main themes of the book. Most policy, certainly public policy, is based upon the idea that people respond as rational individuals, in the sense in which economics uses this word, to incentives. If you fire a shot into the air as a sign to a group of football supporters to disperse, you believe they will be induced to regard the costs of a riot as being too high, and they will disperse. If you threaten to burn someone to death unless he or she changes their religious opinion or ideological belief, you think the negative incentive which such a death entails will be sufficient to achieve your aim, at least as far as most individuals are concerned. And if someone is sufficiently stubborn as to ignore the incentive and to undergo the dreadful ordeal, you think the public spectacle will act as a serious deterrent to others.

* *Daily Telegraph*, 2 April 2011.

In the main examples in this chapter, incentives have not worked in a way which has achieved the desired outcome. But quite often, they do. They even worked up to a point in the religious turmoil of mid-sixteenth-century England. So this approach to policy is not always without merit.

But the impact of networks can be considerably greater than that of incentives alone. If the two operate in conjunction, if we experience the phenomenon of positive linking, the changes to behaviour of a relatively small number of people which incentives might induce can spread across a larger group because of network effects. Equally, however, as we have seen, incentives can be swamped if behaviour across a network surges in a manner different from that intended by the policy makers applying the incentives.

Networks introduce an entirely different dimension into the policy picture. This argument is based on the truism that humans are social creatures. As mentioned earlier, in economic theory, individuals operate like multiple Robinson Crusoes, taking independent, autonomous decisions that are not directly influenced by the decisions or opinions of others. Network theory allows the social dimension of human activity to be taken into account when trying to understand how agents behave, and when thinking through the policy implications of their behaviour.

The examples used to illustrate this so far, those of rioting and being burned at the stake, are rather extreme events which relatively few people will ever encounter. But networks on which people copy or imitate the behaviour of others matter in many, more homely, everyday situations. Think of Crocs. For the few readers who have never come across the term, these are shoes which became very fashionable during the 2000s. Shaped like clogs, the upper material of the shoe is studded with holes. Shoes with holes? These may be perfectly functional in the arid climates of Adelaide or Arizona. But in rainy Seattle or Scotland? Yet the brand proliferated everywhere. And a key reason for the success of Crocs was precisely that they became fashionable in the first place.

A difficult problem, to which we will keep returning during the course of the book, is why something starts to become fashionable at all. Fashion seems to 'emerge' from nowhere in particular. Certainly, many companies, as we will see later in the book, are attempting to make use of the concept of positive linking to try to make their brands fashionable, to promote a cascade of sales. But for the moment we will park this issue safely, and simply note that once a brand, a concept, a way of behaving, starts to become fashionable, it becomes even more so simply because it is fashionable. People decide to buy something, to adopt an opinion, simply because others are doing so. It is this social dimension to choice and decision making which network theory captures. In turn, the increased demand sets up positive feedback on the supply side. More shops stock the item because it is becoming popular, which means that even more people become aware of it.

When network effects are present, people are using rules of behaviour which are quite different from those of the behavioural postulates of economics, in which individuals carefully gather information about alternative choices or courses of action and match them against their own fixed preferences. Instead, they may copy, they may imitate the behaviour of others.

And a key implication of basing your decision, at least in part, simply on what other people are doing or thinking is that your preferences are no longer fixed. Instead, they change and evolve over time, as the impact of people on your various social networks alters your own behaviour. Indeed, the preference of the English soccer followers changed in a matter of seconds once the shot was fired. In terms of fashion, you may be influenced by what you see other people wearing in the street – people you may never see again. In deciding whether to buy into a particular pension scheme, your network here is likely to be a very small number of individuals whose opinion in these matters you trust and value. But in each case, your own private opinion may be influenced and changed by what others are doing.

This is a fundamentally different view of the world from the one in which people are assumed to operate in isolation and to base their decisions on a fixed set of preferences. Even in the latter type of world, the world of standard economic theory, successful policy making might still be difficult. Discovering what people really do want might itself be a considerable challenge. But if their wants, their needs, their desires are fixed, in principle a smart policy maker, whether in the public or corporate sector, can discover them. And the right levers can be pulled, the correct buttons pressed, to change incentives in such a way that the desired outcome of the policy maker can be reliably achieved. In a world in which tastes change in response to the choices and actions of others, this model of policy can simply no longer be relied upon.

The problem for policy makers is even more complicated. Networks can appear in a variety of guises and be activated in a variety of ways. The soccer hooligans were connected by a network of shared interests such as violence and, in case we forget, the game of soccer itself. This is why they bonded together in this particular social network. The choices of individuals in this group on matters such as which brand of trainers to wear, or their opinions on which players should be selected for the team, would be influenced by the choices of others – exactly as they are in the wider, innocuous world of fashion and the purchase of items such as Crocs.

But with the vandals, the shot fired into the air was the signal for a qualitative change in the nature of this network to take place. From being a network of individuals connected in terms of shared interests but nevertheless capable of making decisions as individuals, it became fused into one. A particular mode of behaviour, involving the destruction of property and physical violence, flashed across the entire network. Everyone adopted and became consumed by the same mode of behaviour.

In the example of religious faith, a network of awareness was already in place across England. Many people knew of Cranmer, the Archbishop of Canterbury, and other nationally prominent church leaders such as Latimer and Ridley. At a more parochial level, people would be likely to know of the diocesan bishop, say, or a leading local preacher, even though people on the other side of the country might not have heard of them. So, the population of England was connected on a network which transmitted awareness of the existence of a range of Protestant leaders. The martyrs took the gamble that this pre-existing network, which essentially consisted of one in which information about them was exchanged, could be used by them to influence behaviour and opinions.

But the feature which all these examples have in common, in their widely different contexts and impacts, is that behaviour and opinions can be altered not just by individuals reacting to changes in incentives but directly by what others think, believe or do. The challenge for policy makers in the interconnected, networked world of the twenty-first century is to harness this positive power of networks and to use them in conjunction with incentives. We need networks both to make sense of the policy terrain and to design more effective policies, We need positive linking.

2

'Up to a point, Lord Copper!'

We have seen a range of examples, from harmless choices in foot-wear fashion to altogether more dramatic decisions on ideology, in which the effects of networks were much more powerful than those of incentives. One purpose of this chapter is to redress the balance somewhat by providing diverse examples of incentives having clear and identifiable impacts on behaviour. Successful policy making in the twenty-first century requires an understanding of *both* networks and incentives, a point which is illustrated again as we move through the chapter.

Incentives do matter. This is the one great insight into human behaviour which economics provides, an insight which is supported by an enormous amount of empirical evidence.

I should say immediately that this does not mean that free-market, equilibrium economics is correct. This latter sentence is so important that I will repeat it in bold type. **The fact that agents respond to incentives does not mean that free-market, equilibrium economics is correct.** It is not at all necessary to believe in the whole of the standard behavioural paradigm in economics in order to recognise that incentives matter. Indeed, in the centrally planned economies of the old Soviet bloc, incentives could take non-monetary forms such as social acclaim for meeting the production norm, or for being awarded a medal as a Socialist Hero of Labour.

We will see during the course of the chapter that the use of

'UP TO A POINT, LORD COPPER!'

incentives to achieve aims or targets, perhaps changing tax rates or giving out subsidies or medals, is by no means a panacea for the policy maker. This is the case even when network effects are either weak or absent more or less altogether. Humans are inventive, innovative, and they may very well respond to changes in incentives in ways which are very hard to anticipate.

So incentives work only 'Up to a point, Lord Copper', as the editor of the *Daily Beast* in Evelyn Waugh's novel *Scoop* used to say when his titled proprietor asserted something which was at best only partly true and was often unequivocally wrong.

Sometimes, changing incentives does work out more or less as expected. This is certainly true in a qualitative sense, even if the exact quantitative predictions are not borne out. If Coca-Cola, say, puts the price of its products up and nothing else changes at the same time, to state that the sales of Pepsi will probably go up is perhaps merely stating the obvious. Agents who like this sort of fizzy drink are given an incentive to buy less Coca-Cola: its price has gone up. But incentives have been used in less expected ways, sometimes by the most unexpected people.

For example in 2003 Ken Livingstone, as mayor of London, introduced the 'congestion charge', a tax on vehicles entering central London during the day, in an effort to solve the problem of traffic congestion. Even the mayor's worst enemies could scarcely accuse him of being a gung-ho free-market economist. His political stance has always been firmly on the left.

Nevertheless, in a politically bold move, Livingstone attempted to deal with the traffic problem in a major world city by the use of incentives. There was great uncertainty in advance, which persisted even during the early months of the actual operation of the scheme, about how agents – motorists in this case – would respond. Many different forecasts were made. But the tax has worked *qualitatively* exactly as one would expect. Traffic flows

37

into Central London are lower than they would have been without this charge. In other words, faced with an additional cost of
driving into central London, some motorists have decided either
to reduce their visits to the area, or to use alternative means of
transport.

Livingstone made use of the reaction of agents to incentives to
achieve a desirable social goal. It does not mean he had become
a convert to the political ideology of Mrs Thatcher. Nor does it
necessarily mean that motorists carried out a rational analysis of
the costs and benefits of the scheme. But it worked.

Much more generally, the economics of the mainstream works
up to a point exactly *because* it incorporates this fundamental
insight into human behaviour, that changes to incentives alter
it. But such insight is perceived as through a glass darkly. Except
in certain limited circumstances, the rest of its theoretical con-
structs are at best shaky and often plain wrong. The core model
of standard economics assumes that an agent gathers all the available information relevant to a decision, and is then able to process it in a way which enables the agent to arrive at the very best
decision possible, given the tastes and preferences of the agent.
These latter are assumed to be fixed and cannot be influenced by
what other agents do. All theories are approximations to reality.
The question is always: how good are the approximations?

Policy based on the use of incentives is mistrusted by many
people, precisely because it has the image of being derived from
the highly mathematical abstractions of economic theory. So,
before going on to give examples of unintended consequences of
changes in incentives, it is worth considering briefly whether the
use of abstraction and maths can be justified in analysing social
and economic problems. We are, after all, dealing with human
behaviour.

I have no problem at all with abstraction, though I am mind-

38

ful of the fact that this needs to be justified. Many scientific theories are highly abstract, far removed from everyday life. But this abstract quality is the very nature of the beast. From the myriad complicated details which surround many situations in reality, we are trying to distil a few key factors and to describe how their interactions help us to understand what is going on. Ideally, we would like this theoretical account not just to shed light on one particular problem, but to be capable of generalisation across many situations.

Ultimately of course, any theory, if it is to be regarded as being truly scientific, has to be tested by empirical evidence. It must be judged not by its abstract beauty, but by its ability to explain messy reality. This is how, for example, we are able to dismiss astrology as not being a science. For a number of years at the start of my career I was an economic forecaster, attempting to predict the future course of the British economy. Despite being equipped with the latest red-hot developments in both economic and statistical theory, the ability to achieve any degree of systematic accuracy proved elusive. Growing disillusioned with the whole process, for a short time I carried out a comparison between the accuracy of my macroeconomic forecasts and those of my daily horoscope. If anything, the latter had the edge!

But theoretical abstraction is both desirable and necessary for any sort of progress to be made in understanding the world, whether in the natural, biological or social sciences. So the abstract nature of much economic theory does not of itself make it a legitimate object of criticism. Even if the theory is misleading, or even plain wrong in many situations, this is not because of its abstract nature. It is because the assumptions, the simplifications which are needed to have any sort of theory, are not supported by the evidence.

*

It is rather harder to justify the use of mathematics in economics, especially the specific sort favoured by economists. Maths is pervasive in economics. It is an integral part of the self-image of the discipline. And it serves as a very distinct barrier to entry, a very clear 'Keep Out!' message to anyone who does not feel comfortable in this area. Given its importance, a fairly lengthy detour is warranted before we move back to an explicit discussion of incentives, the main theme of the chapter.

Maths can in fact be very useful, provided that we think of it as just a tool. Economists often make clever maths an end in itself, and in doing so overlook the fact that we are trying to understand and explain what happens in the real world. Elegant maths also often leads economists to make the mistake of confusing the model with reality. It is a tool which can assist logical thinking. It's like another language. It can help us find our way around and serve as a medium of communication amongst people discussing the same subject.

Here is a cautionary tale of how maths has become fetishised at the very highest levels of the discipline of economics. A friend of mine teaches economics at the University of Cambridge in England. Fairly recently, she had a first-year student who was very good indeed at maths. So much so that he complained there simply was not enough of it in his course. For his second year, he was sent on an exchange to the other Cambridge, to the Massachusetts Institute of Technology. Emails of an increasingly desperate nature began to whizz back to my friend across the Atlantic. The final one said simply: 'Help! Please let me back home. There isn't any economics in this course. It's *all* maths.'

Things are not quite as bad as that in most places, but the use of maths has become pervasive in economics. Just for the record, at the right time and amongst consenting adults, I, too, use maths extensively, albeit of a different kind from that which pervades economics. You can tell I am an economist myself when I say, on the one hand there are good reasons for the use of maths in economics, and on the other hand there are bad ones. So far, mainly the bad ones have prevailed.

It would not matter very much if economics was not taken so seriously by policy makers. Hardly anyone bothers about some of the lunacies in literary theory, for example. But economics matters. Why is it that maths came to be so pervasive in economics, when so much was achieved without it? The worst reason is that the use of maths makes economists feel that they are proper scientists. They suffer from deep physics-envy. Ironically, economists seem to envy the classical physics of a century and more ago. In many ways, physics itself has moved on to incorporate network-based ideas.

Physicists have to use maths – try doing quantum physics in words. And they are real scientists, who really have explained how lots of things really do work. So if we use maths, that makes us real scientists, does it not? The logical error in this last sentence is pretty obvious. But it does not stop the inner glow of satisfaction that most economists feel when they cover the page in mathematical symbols.

There is a more serious and more damaging reason why maths, or at least a particular kind of maths, is used in economics. This is inextricably linked with the concept of Rational Economic Man. In essence, as noted in the previous chapter, economics is a theory about how individuals behave. And in the standard theory, it is not just that people are assumed to be self-interested. Rational Economic Man acts like some sort of super-computer, always gathering every single bit of information which is relevant to a decision, and then making the best possible decision out of all the available options. Not just a good decision, or even a very good one, but the best. The 'optimal' decision.

Now there is a whole branch of maths devoted to 'optimal' solutions. This is differential calculus, which many readers will have come across at school. It is the ideal tool for a theory which says that individuals behave in a way which is optimal for them, given their tastes and preferences. So if you eat junk food and weigh 300 pounds as a result, or if you drink heavily and destroy your liver, or if you smoke and get cancer, if you riot when the police open fire, that is your choice. You must have been making what you believed to be the best possible lifestyle choice for you, and calculus can prove this.

This is still the basis for a lot of the economics which is taught today. Yet, paradoxically, it has been precisely the use of maths within economics itself which has exposed fundamental problems at the very heart of the model of the Rational Economic Man theory of behaviour.

Working out the full implications of these behavioural postulates proved an exceptionally demanding scientific task, which took a century to complete. By the mid-1970s, this programme of research was eventually finalised. There is nothing left to discover. It is a marvellous intellectual construct, but it turns out to be a scientifically empty box. It has no testable implications. In other words, there is no empirical test we can use with which the theory could be refuted. Such tests might be very hard to devise, but any true science has to be capable of being refuted empirically.

*

We might pause and offer some postulates which might be capable of being refuted by evidence. For example the familiar diagrams of basic Economics 1.01 show downward-sloping demand curves. This simply means that a lower amount of a product is demanded if its price goes up. A higher price means less sales by volume, and if we plot such a relationship in a simple chart, the relationship between sales and price will slope downwards. Figure 5.2, on page 151, illustrates the point. But we cannot deduce logically from the theory of Rational Economic Man behaviour that this key statement, widely used by economists, is true.

This result was established through pages of intricate maths. A translation into English could be carried out, but a full explanation would take many chapters. However, a useful insight into the proposition that we cannot deduce theoretically that market demand curves slope downwards, even if the demand curves for every single agent do slope down, is as follows. Suppose the price of a product is increased. The volume of sales will go down. The value, which is just price multiplied by quantity, may either rise or fall depending on how much the volume changes compared to the increase in price. If a 10 per cent price increase only reduces volume sales by 1 per cent, the value of sales will rise, but if it cuts volume by 20 per cent, value will fall as well. Either way, the income of the company making the product is changed. If the market for the product is tiny compared to the economy as a whole, we can reasonably stop our analysis of the implications of the price rise here.

But suppose we are thinking about the demand for labour, about how many workers firms want to employ. This market – or collection of markets – is typically enormous relative to any national economy. A persistent theme in economic discourse, and indeed policy, is the need to 'price people back into work'. In other words, to reduce wages so that more people are employed. A cynic might note in passing that the argument is usually applied to less skilled people, and few in the financial services sector have been suggesting that bankers take a salary cut in order to price them back into work after the financial crisis. However, this is indeed a point made in passing.

Suppose further that, by some means, a government succeeded

in cutting wages to increase the demand for labour. The potential problem here is that wages are not just the 'price' of employing someone to the company, they represent spending power across the whole range of goods and services. So a reduction in the price of labour might lead to *fewer* people being employed because consumption by the workers, the total amount they spend, may decline too. On the one hand, labour has become cheaper. On the other, workers have less to spend throughout the economy as a whole, so the demand for many products may fall, and fewer workers are needed to produce them. Of course, there could be an offsetting effect if firms spent any increase in profits resulting from the wage cut. This account, however, does give some insight into why it is not possible theoretically to prove that 'demand curves slope downwards'.

Despite the theoretical indeterminacy of the core model of conventional economics on this point, the idea that labour is too expensive, that it should be priced back into work, that real wages should be reduced, was a key ideological theme of the 1990s and 2000s in many Western countries.

Of course, the reality is that if the price of a product or service is increased then usually, but not always, its sales fall. In practice, and certainly when added up across any particular company, salaries cannot be significantly higher for a long period of time than the value of the contribution of the workforce. If they are, the firm will eventually fail. But these are empirical observations, which seem generally valid but which do not obtain on every occasion. The point about the theoretical model is that even if we observed the opposite, that, say, when the price of a product rises its sales increase, this would not be a refutation of the theory, because the theory allows the relationship between price and sales to take any shape whatsoever.

So, paradoxically, the use of maths in economic theory, and

more precisely the type of maths preferred by economists themselves, has provided some very powerful results which do much to undermine many of the policy-related claims it makes.

*

In practice, of course, returning directly to the main topic of the chapter, incentives often matter. To stress again, this statement does not necessarily imply that the theory of Rational Economic Man is correct. Further, when we speak of 'incentives', the usual concepts associated with the word are things like pay and prices. You may be offered a bonus at work in order to try and incentivise you, either to work hard or to stay loyal. A government may increase tax on fuel, say, to try to get people to use less of it, to respond to the negative incentive of the higher price. Yet as we have seen, incentives may often appear in an unconventional form, such as the 'price' of being burned to death.

It is not just that incentives may appear in odd or unexpected guises. Their effects may be hard to anticipate. This does not mean that they are not working at all. Nor does it mean that we have simply just not done enough careful statistical analysis of behaviour to be able to say that if, for example, we put the price of Coca-Cola up by 10p or 10 cents, sales will fall by x or y per cent. Rather, it can mean that agents *do* respond to changes in incentives, but the ways in which they respond may not be anticipated by the policy maker. Or it can mean that the incentives work in the way they were intended to, but that the wider consequences of this are not foreseen.

On the latter point, a few years ago, my wife telephoned to make an appointment with a doctor at our local National Health Service surgery. It was not urgent. Previously, she had been able to make such an appointment at her convenience several days in advance. But the rules had changed. Appointments could now only be made on the same day that the request was made. The government had brought in a target that a high percentage of patients had to be seen on the same day that they contacted the doctor. On the face of it, a perfectly desirable aim. Doctors received payments which were conditional on the target being met, so they were incentivised to do so.

But this led to great inconvenience. Working people cannot always guarantee to be free on any particular day, and even when they can, others may get in before them and take the slots. When she finally managed to get to see the doctor, many fruitless phone calls and over a week later, she raised this policy change with him. The doctor was most apologetic. It was not his fault. His funding depended in part on him meeting the government's new target. The doctor was responding to incentives in a perfectly sensible way from his point of view. The unintended consequence was that his unfortunate patients experienced considerable inconvenience in no longer being able to make appointments to suit their schedules.

The limitations of what we might term the 'clever regulator' approach to policy, using incentives to achieve specific targets, were also present during the UN Climate Change Conference in Copenhagen in 2009. This time, the authorities failed to grasp how agents might respond innovatively and imaginatively to the incentives put in place. The city's mayor, Ritt Bjerregaard, and the city council wanted to curb prostitution during the conference. They sent postcards to hotels and delegates to the conference urging them not to patronise the city's sex workers. The delegates were exhorted to 'Be sustainable – don't buy sex'. The hotels themselves were admonished: 'Dear hotel owner, we would like to urge you not to arrange contacts between hotel guests and prostitutes'. Here was a clear set of incentives placed in front of agents. The incentives were perhaps particularly strong for the hotels, for it is not in the interests of service providers such as these to incur the potential wrath of the local authority, equipped as it is with all kinds of powers which can make their lives difficult.

The response of the prostitutes shows how inventive humans can be. Members of the Sex Workers Interest Group simply offered free sex to anyone who could produce both their delegate credentials to the UN conference and one of the notices sent out by the mayor. Their incentive to maintain their business was sufficiently strong for them to introduce this innovative marketing arrangement. The choice they faced was between a much reduced income if the mayor's strategy was complied with, and a normal income reduced by the occasional free service.

A much more detailed example of unintended consequences, or, more precisely, consequences which are very hard to foresee because of the innovative responses of agents, is as follows. Jérôme Adda and Francesca Cornaglia of University College London published a study in 2006 in the top-ranking *American Economic Review*. The potentially detrimental effect of nicotine on smokers' health is well established. In recent decades, most Western governments have attempted to reduce cigarette consumption, and an important way of trying to achieve this aim has been by increasing excise duties – taxes – on cigarettes to make them more expensive.

The policies have undoubtedly been successful. A number of detailed academic studies have shown a distinct correlation between higher prices and reduced cigarette consumption. The initial impact of a tax increase tends to be diluted over time because of the addictive nature of the product, but it nevertheless persists.

In this context, we might usefully note that social networks have reinforced the impact of incentives, and in particular by their influence in persuading people to stop smoking altogether. The Framingham Heart Study is a unique database, monitoring the health of individuals over many decades in the eponymous town in Massachusetts. It is a rich source for medical research. But it also provides material for social scientists. Unusually for such medical surveys, the study contains information not just on the individuals but on their family and friends.

Nicholas Christakis and James Fowler of Harvard University analysed the data from a network perspective, publishing their findings in the *New England Journal of Medicine* in 2008. Their results were striking. The cessation of smoking by a co-worker in a small company decreased a person's chances of smoking by 34 per cent. If a friend gave up, the person was 36 per cent less likely to smoke, and the chances were 59 per cent less if a spouse stopped smoking. Their study does not identify the separate impact of incentives on individuals, such as price increase or the public information provided on the health risks of smoking. But once an individual makes the decision to stop, for whatever reason, the effect is potentially transmitted through his or her social networks – family, friends, work colleagues – to others, who might stop simply because of the example which has been set to them. So networks can operate *with* incentives, to reinforce and magnify the initial impact of the latter.

To return, however, to Adda and Cornaglia and the potential difficulties of anticipating the effects of changes in incentives. Their article is based on the so-called 'rational theory of addiction'. Full of heavy-duty maths, it is replete with phrases such as 'we assume a quadratic utility function' and 'the proof requires a second-order Taylor approximation'. But, to reflect on points made earlier in the chapter, we really do not need the assumption of rational behaviour at all. The empirical results of the study, based upon careful statistical analysis of the data, are clear.

They used the American National Health and Nutrition Examination Survey, a database of some 20,000 people across the United States, which contains information on the number of cigarettes smoked and their nicotine, tar and carbon monoxide concentration. Tax rates on cigarettes vary across states, providing plenty of variation with which to estimate their impact on behaviour. In common with many other studies, Adda and Cornaglia found that the higher the tax rate, the fewer cigarettes were smoked. So far, good and entirely expected news for the health-promoting policy maker. But they discovered that higher tax rates led smokers to switch to brands with a higher tar and nicotine yield. This was not in itself a novel finding, though it increased the credibility of the two previous research papers which had previously reported it. A large number of papers had been written on the impact of taxes and prices on the number of cigarettes smoked, but only two on this switching behaviour, so it was valuable confirmation of this effect, though nonetheless worrying.

The real originality of the research was the discovery that smokers also increased the intensity of their smoking by extracting more nicotine per cigarette, regardless of the brand which was consumed. Smokers become more inclined to smoke the cigarette right to the end, behaviour which not only increases tar and nicotine consumption, but also leads the smoker to inhale more dangerous chemicals, which in turn has been shown to cause cancer deeper into the lung. So, yes, higher taxes do reduce the sales of cigarettes. Incentives work as expected. But at the same time, smokers compensate by both switching to brands containing more tar and nicotine, and by consuming cigarettes in ways which are more dangerous to their health.

So, traffic in central London, the response of professional doctors to changes in payment structures, the supply of sexual services, nicotine consumption – a disparate range of circumstances in which incentives have altered behaviour. Sometimes in ways which the policy maker did not foresee and did not, in hindsight, desire. But incentives undoubtedly mattered.

Network effects have also been demonstrated to be important, reinforcing the impact of incentives, in at least one of these examples. And in general, as we saw in the opening chapter, in most real-world social and economic situations, we need to understand the potentially subtle and powerful interplay between the effects of incentives and the effects of networks.

*

So far in this chapter we have focused deliberately on incentives, on some of the strengths and weaknesses of the traditional tool of policy makers. It is time for a shift of gear. Time to discuss at some length an important policy area where both network effects and incentives are at work.

The complex relationship between incentives and networks to which I refer is that within the criminal justice system. A perennial question for policy makers is: does prison work? It is at this point that we put on our detached philosopher's hat and ask in turn what precisely this question means.

Even in the United States, where the rate of incarceration per 100,000 inhabitants is five or six times the average in the rest of the developed world, it is hard to get sent to prison. There are occasional highly publicised stories in which individuals leading hitherto blameless lives suddenly in a fit of rage murder their spouse, or even go on a killing spree. But the vast majority of people who get sent to prison have already had fairly extensive experience of crime and the criminal justice system. They are steeped in the culture of crime. Superb television series such as *The Sopranos* or *The Wire* hold our attention through the quality of their scripts and acting. But they also succeed because they are realistic. Almost like Dante's Circles of Hell, there are concentric rings of individuals, from the hard-core gang leaders out through people only peripherally engaged with them. But they are all engaged with the process of crime. And not surprisingly, once they become engaged in this way through their social networks, some of them end up in prison.

One thing we have observed empirically about crime is that once a person has been in prison, he (or, far less frequently, she) is very likely to commit a crime again. Recidivism rates are high, even though a vast range of policies have been tried in an attempt to get the rate down. In this sense, prison does not work.

Despite stories in the popular press about how criminals live a life of luxury in jail, in reality a prison term remains an unpleasant experience for most of those incarcerated. Apart from the loss of liberty, crime within prisons is often rampant, and many inmates live in fear of physical violence. Tom Wolfe's description in *A Man in Full* of the Californian prison in which one of the book's more sympathetic characters, Conrad Hensley, is incarcerated is awful, brutal – and entirely accurate. Yet the experience seems to provide scant deterrence against reoffending. Once someone is sent to prison, the influence of social networks on individual behaviour appears to dominate that of incentives. The social and cultural milieus of hard-core criminals, the social networks in which they are embedded, are ones in which crime is itself the norm.

A key fact is that most crime is committed by young men possessing little money or intelligence and few skills. And there does appear to be something inherently implausible about the idea that such individuals assess all the available information and choose the 'optimal' decision when they are contemplating breaking into a car or thinking about punching someone in a bar.

The standard response by economists to such points is to invoke the 'as if' argument. In other words, whilst it may not appear that agents go through the process of finding optimal decisions, they behave 'as if' they do. There are layers of subtleties to this argument which need not delay us. But even the simple statement of the point is not as foolish as it might first appear. Very few of us know how to solve the difficult non-linear differential equations which describe the flight of a thrown ball, yet most of us can predict its path well enough to catch one. It is 'as if' we had done the maths.

But, to repeat the point made at the start of the chapter, we do not need to invoke the idea that agents are responding in some optimal sense to incentives. People may be short-sighted in terms of the decisions they make, they may consistently make decisions whose outcomes go against their own self-interest. Yet they are still changing their behaviour in response to changes in incentives.

This is a point which most economists find hard to accept. Surely, the mainstream 'rational' agent argument goes, people do not necessarily always make the optimal choice, but rather over time they gradually learn to avoid decisions which are not in their own interests. This raises an issue of great importance to which the whole of the next chapter is devoted. In essence, in many situations, the best choice can *never* be identified, even after the event, no matter how smart we may be.

For now, however, in the spirit of empiricism, we simply note that, for most people, crime does not pay. Most of the young men who spend their days steeped in petty crime would actually be better off in straight money terms in low-paid legitimate employment. The proceeds from most crimes are very small. Criminals often act impulsively, paying less regard to the potential costs to them of committing a crime than the objective evidence indicates they should. For example, as known criminals in a locality, they have to endure the stress of frequent visits from the police, their own records making them natural suspects. So the benefits from crime are not as high as those from a regular job, and the costs – the stress of being known criminals, the court appearances, the fines, the frequent prison sentences – are much higher. Yet individuals persistently choose to follow a criminal life. In part, this can be explained by the culture in which they become involved, the social network of crime. But in part it can only be said that they are making a constant stream of decisions which, from the point of view of economic theory, are irrational, which go against their own self-interest.

However, this certainly does not mean that criminals fail to respond to incentives. True, they make decisions which, in terms of their own self-interest, are often not very sensible. True, they do not necessarily respond exactly along the lines of the theory of Rational Economic Man. But their behaviour can nevertheless be influenced by the various positive and negative incentives which criminals face.

An example of positive incentives occurred in April 1999, when Britain introduced for the first time a minimum wage, which provided pay increases for a large number of low-paid workers. Two London-based economists, Kirstine Hansen and Steve Machin, carried out a careful and very sophisticated statistical analysis of its impact across the forty-three police-force areas of the UK. They concluded unequivocally that 'altering wage incentives can affect crime and therefore that there exists a link between crime and the low wage labour market'. By making the alternative option of regular employment, albeit at the minimum wage, more lucrative, some potential criminals were incentivised to choose this rather than to 'earn' their living from crime. This does not mean that they had suddenly become rational agents in the economic sense of the term, able to assess costs and benefits more effectively. It simply means that incentives had changed and, however imperfectly, some agents responded.

*

Steve Levitt is famous for his blockbuster book *Freakonomics*. But he is also an extremely distinguished economist, winner in 2003 of the John Bates Clark Medal awarded to 'that American economist under the age of forty who is adjudged to have made a significant contribution to economic thought and knowledge'. One of his areas of interest is crime. And the discussion of why crime has fallen in America in his book is based in turn on an article he wrote in the prestigious *Journal of Economic Perspectives*, which in turn is based on a large number of technical academic articles.

One of his conclusions has struck a notable chord with many people, quite possibly because of the rather startling and unexpected nature of the topic: that one reason for crime falling sharply is a rise in the number of abortions. As noted, most crime is committed by poor, unskilled young men, and more abortions in their social group means that there are fewer of them around to commit crime.

But Levitt also concludes that prison is in part responsible for the dramatic reductions in crime, a finding often conveniently forgotten. Both Britain and America saw large increases in the prison population starting around twenty years ago, and in both countries there have been sharp subsequent falls in crime. Of course, simple correlation such as this does not prove causation, but Levitt's conclusions are based upon highly sophisticated statistical studies which readily encompass such issues.

One obvious factor is that people in prison cannot commit further crimes – at least against society in general. So, by simple arithmetic, a bigger prison population means smaller crime figures. But the more important impact arises from the deterrent effect, not on those who are actually serving sentences, but on those at liberty who are contemplating breaking the law.

A key step seems to be moving from a situation in which a young man has not committed a crime, to one in which he has. Although most crime is committed by poor, unskilled men, most such individuals remain law-abiding citizens. They will probably know who the criminals are in their neighbourhood, and may even socialise with them. But the main influence of their particular social networks, the main impact on what they regard as normal behaviour, remains people like themselves who live their lives within the law.

Of course, carrying out a single criminal act does not immediately lead to the destruction of an individual's existing social networks and to his re-embedding into a group of hardened career criminals. But it may begin to alter these relationships.

An elusive goal for criminologists is to identify individuals who are more likely to become prolific criminals (at some point in their lives) than others. The group of prolific criminals is usually thought to be some 5 or 10 per cent of the total population of offenders. Some progress has been made. Being born into a family where most members are criminals increases this probability substantially. And it is now clear that boys raised by single-parent, never-married mothers also exhibit a higher probability of being involved in crime than others with different family backgrounds. I should stress that it is not the case that boys from such backgrounds automatically become criminals. Indeed, most grow up to be perfectly respectable members of society. Nonetheless, the probability of them turning to crime is distinctly higher than it is for the population of boys as a whole, even taking into account factors such as family income. In general, the vast majority of offenders share the characteristics of the persistent offenders, making prior identification very difficult. Many such offenders abandon crime during their twenties and become productive, taxpaying citizens.

It is, of course, neither practical nor acceptable to incarcerate boys from criminal families as soon as they reach puberty, still less every boy from a poor, single-parent family. Could we not instead attempt to identify the much smaller number of those who are likely to commit large numbers of crimes and devote resources diverting them from such a path before it is too late?

If we could do so, there would be a double impact: first, fewer criminals would mean fewer crimes; second, the influence of criminality as a social norm amongst their peers would be weakened, since individuals known to have committed large numbers of crimes undoubtedly attract attention and gain influence in such social circles.

*

Surprisingly little systematic work has been done on the number of crimes committed by individuals, but there are two well-established databases which record criminality amongst a group of individuals over time. The first, the Cambridge Study in Delinquent Development, is a prospective longitudinal survey of 411 males in a working-class area of north London. Data collection began in 1961–2. The second, the Pittsburgh Youth Study, began in 1986 with a random sample of boys in the first, fourth, and seventh grades of the Pittsburgh public school system. The sample contains approximately 500 boys at each grade level, for a total of 1,517 boys.

The Cambridge data relates to the number of convictions for each boy over a period spanning the mid-1960s and 1970s. The Pittsburgh data describes self-reported acts of delinquency over short time intervals beginning in the late 1980s. In other words, the studies differ both in their time coverage and in the fact that the Cambridge study is based on convictions whilst the Pittsburgh one utilises self-reporting.

Despite these differences, there is a remarkable similarity between the two in the statistical distribution of the number of crimes committed. The 'statistical distribution' in this context describes how many individuals in each database commit (or record) zero crime, how many commit just one, how many commit two, and so on.*

There are two striking features of the results. First, a much better description of the number of crimes committed by individuals is given if we segment the number into two separate groups than if we analyse them all together. Specifically, the groups are 'the numbers who commit zero crimes' and 'the numbers who commit any crime'. In other words, the description of the data when the number of boys committing or reporting zero crimes are excluded is different from that when they are included.

Second, once this distinction is made, there is no 'typical' number of crimes which an individual commits. Once a boy has moved from committing no crime to committing just one crime, the scale of his resultant career in crime could turn out to be small, medium or large. Moreover, the number of crimes which any individual does in fact commit can be thought of as the outcome of a purely random process.

We see again, incidentally, the concept of 'robust and fragile'

---

* The actual analysis relies on a number of mathematical concepts which would take considerable time to describe in words. For those interested in the details, I have published the analysis as 'Scaling Behaviour in the Number of Criminal Acts Committed by Individuals', *Journal of Statistical Mechanics: Theory and Experiment*, July 2004. It may be thought unusual that a statistical physics journal would be interested in this analysis, but the statistical distribution which is identified is one of general interest to this particular research community.

networks introduced in the opening chapter, albeit in a slightly different guise. Here we have a network, in this case a population of young men, living on the same public housing scheme perhaps, who have not yet committed a crime. For whatever reason, one of them carries out a criminal act. Most of the time, he will never go on to commit more than a handful of crimes. Occasionally, he will graduate to a life involving numerous criminal acts spread over a period of years. The network is robust in the sense that most of the people in it carry out either no crimes at all or just a small number. And it is fragile because a small number do go on to be career criminals.

These abstract concepts have two important practical implications. First, the fact that the number of crimes committed by an individual is compatible with the outcome of a purely random process means that it is not possible to identify in advance, once a crime has been committed, how many crimes that individual will go on to commit. So we cannot hope to target *in advance* those boys who will have a highly prolific career in crime, and who may therefore exercise a strong influence over the behaviour of their peers. We may, as discussed above, be able to go some way in identifying those who are more likely to make the first crucial step from zero to one crime, but we cannot then go on to separate those who will commit many more crimes from those whose criminal career will involve only a small number.

The second is that the crucial step is indeed to make the transition from being law abiding to carrying out the first criminal act. In terms of the numbers of individuals committing different numbers of crimes, more commit just one crime than commit two, more commit two than three, and so on. But the largest and most important distinction by far is between zero and one.

Here is where we see an interplay between incentives and networks. Once a young man makes the initial transition to crime, his perception of himself starts to alter, as does the perception others have of him. He becomes potentially less acceptable to the members of his various social networks who do not commit crime, and, conversely, more in tune with the social values of those who do. He may himself accelerate the process of potential change, depending on how much his own self-image is altered as a result of his actions. His identity changes. Once a young man has carried out a few crimes, he is by no means predestined to become a career criminal. Indeed, most do not: the effect of the non-criminal social networks in which he has been involved as a non-criminal often draw a young man back into a law-abiding life.

Incentives, whether positive or negative, may influence either the crucial initial decision to commit the first crime, or, later, the decision to withdraw from a life of crime. But the influence of peer pressure, peer acceptance, the gradual increase in the relative importance of copying or imitative behaviour compared to that of incentives, increases the more a young man associates with criminals, and there is a chance that a criminal career has been born.

It is this which provides the intellectual basis for successful policies of containing crime. It is the positive use of networks of attitudes amongst the 'at risk' group, the search for the triggers which will generate positive linking across these networks, which will keep crime down. There is no check list of policies, each 'rationally' evaluated by teams of economists, which will guarantee success. Rather, it is the much more subtle concepts of social norms, of what constitutes reasonable behaviour in the relevant peer groups, which is the key. This is hard to achieve, not least because there is no readily specified tick-box approach to the problem. But positive linking has the potential to create massive changes for the better.

Yet a nagging question runs through this whole discussion of crime. Most criminals have backgrounds of poverty, they have low levels of conventional skills, and are often barely literate or numerate. It seems implausible that they behave as 'rational' economic agents, gathering information and meticulously processing it in order to arrive at the best possible decision, given their own tastes and preferences. Indeed, we have argued that this is *not* how we need to see them as behaving. Agents can still react to incentives even though they are not following the behavioural precepts of conventional economic theory.

The question is: does economics have anything to say about behavioural patterns which do not square with its core theoretical assumptions? Do we simply dismiss such behaviour as 'irrational', or is there something more useful we can say?

The concept of the rational agent does indeed remain very much alive and well within economics. But, as it happens, in recent decades, a whole new empirically driven field has developed within the subject itself, one which poses challenges to the mainstream view of how the world operates. This is known as 'behavioural economics', and it is to this which we now turn.

3

The Shoulders of Giants: Simon and Keynes

An innocuous pastime of sports fans is to discuss and debate an 'All-Time Team', the best players of all time, whether in the sport as a whole, be it soccer, basketball or whatever, or for the team they support.

One person who is at or very near the top of both my own and many other people's lists of the greatest ever team of social scientists is Herbert Simon. Born in 1916 to parents who moved to America at the start of the twentieth century, Simon's initial education and career was at the University of Chicago. In 1949 he became Professor of Industrial Management at Carnegie Mellon, and continued to teach in various departments until his death in 2001. He was awarded the most prestigious prizes in several disciplines. He received the Turing Award, named after the great Alan Turing, the father of modern computers, in 1975 for his contributions to artificial intelligence and the psychology of human cognition, and in 1978 he won the Nobel Prize in economics 'for his pioneering research into the decision-making process within economic organisations'. In 1993 the American Psychological Association conferred on him their Award for Outstanding Lifetime Contributions to Psychology.

He carried out outstanding original research in cognitive psychology, cognitive science, computer science, public administration, economics, management, the philosophy of science, sociology and political science. Simon often anticipated developments in a particular field years, and sometimes decades, ahead of the mainstream. In 1957 he predicted that computers would outplay the most proficient humans at chess within ten years. In this case he was wrong, but only on his time scale, for this did not happen until the 1990s. But remember that this was at a time when computing was in its infancy, when machines the size of a small house were nowhere near as powerful as the most basic laptop, or even smartphone, of today, and ideas such as this were the stuff of science fiction.

Perhaps his most significant intellectual contribution was in creating virtually single-handedly the field of what is today known as behavioural economics. Unlike most economists, certainly then and still to a large extent today, Simon took an active interest in how firms actually behaved. Companies are the foundation of the prosperity of the developed economies, and they are by a very considerable margin the most successful organisational innovation in the economic world over the past 200 years.

Simon's exceptionally innovative work on behavioural economics was carried out in the 1950s and, as we shall see, its main message is still very far from being accepted by the mainstream economics profession. We might reasonably ask, if his theories are as brilliant as I am painting them to be, why are these now not the accepted wisdom in economics? As we will shortly see, much of his work has indeed been incorporated in various ways, but his key, revolutionary message has been safely neutered by the mainstream. There are many reasons for this, but economics has a bad track record of either not rejecting its core theories when they are brought into serious doubt, or not accepting scientifically superior theories. So before examining Simon's theories, it is useful to spend a bit of time examining a key aspect of firm behaviour in

economic theory, and showing how this had already been rejected empirically when Simon was writing, but that economists persisted with their incorrect assumptions nevertheless.

In much of economic theory, size is not just irrelevant but
a definite handicap. A concept known as 'diminishing returns'
prevails, so that as more labour and more capital are used in the
process of producing any particular good or service, the extra
output obtained from each additional unit inputted into the
process eventually falls. This theoretical assumption is not made
on the basis of empirical evidence. It is made because it makes the
maths, hard enough as they are, much more tractable, easier to
handle analytically. But in the real world, increasing returns are
widespread. Companies often gain distinct advantages through
utilising economies of scope and scale in their operations. As they
get bigger, unit costs often fall, not rise.

Ironically, at the very same time that the concept of diminishing marginal returns was capturing the academic discipline of
economics, the United States was moving towards world economic dominance by exploiting the unprecedented and massive
increasing returns to scale of production and distribution which
its rapidly expanding economy permitted. In other words, by
taking advantage of the benefits of being big.

A large number of the companies which were to become
household names in the United States in the twentieth century grew enormously and established their market strengths in
the final decades of the nineteenth. Quaker Oats, Campbell's
Soup, Heinz, Procter & Gamble, Schlitz and Anheuser Brewing,
Eastman Kodak, American Telephone and Telegraph, Singer,
Westinghouse, Union Carbide – all these are examples of companies which took advantage of increasing returns in production
and distribution at that time to establish themselves on a national, and in some cases international, scale.

In 1870 the population of America was 39 million, not much more than that of Britain at 31 million. American income per head of the population was around 80 per cent of the British level, which was then the highest in the world. The combination of a larger population and a lower income per head meant that the size of the American domestic market was virtually identical to that of the British.

By the onset of the First World War, this had changed dramatically. America had overtaken Britain as the leading economy in terms of income per head. From being 20 per cent below the British figure in 1870, America was by then 20 per cent above. And whilst the population of Britain had grown by only 14 million over this period, in America the growth was 58 million. Indeed, the growth in the American population was of itself much bigger than the total size of Britain's population.

From a position of equality in 1870, by 1913 the American domestic market was over two and a half times that of its main European rival. Such a market was of a size entirely without precedent in world history. It offered tremendous opportunities for the exploitation of increasing returns to scale. Companies during this period found, contrary to the precepts of marginal economics, of which they were blissfully ignorant, that the bigger the scale of operations, the more could be produced, and the more profit could be made from the additional marginal unit of production.

Advances in technology made these leaps in economic progress possible, which in turn provided the finances not just for further investment in plant and machinery but for yet more research and development. Such positive feedback placed the Western economies, and particularly that of America, on a virtuous circle of growth.

So by the time Simon began his research in the 1940s, giant corporations had been in existence for half a century. The assumption in economic theory of diminishing returns in the process of production in these outfits had simply not been not borne out in practice. But, as already noted, economists continued to work with such assumptions in the core models of their theory.*

Simon's interest was in the validity of another concept, arguably even closer to the very core of economic theory, namely that agents – whether individuals, firms or governments – behave in a so-called 'rational' way. How did firms really behave, how did they actually make decisions? Remember that his initial Chair at Carnegie Mellon was not in economics, but in industrial management.

He published his reflections over the years on these matters in 1955 in a remarkable article in the *Quarterly Journal of Economics*, entitled 'A Behavioral Model of Rational Choice'. The article itself is theoretical, but throughout the paper Simon makes explicit the fact that his choices of assumptions are based upon what he considers to be empirical evidence which is both sound and extensive. This truly brilliant article, as already noted the basis for the whole field of behavioural economics, is worth quoting at some length.

Simon begins the paper with what by now will be familiar material:

Traditional economic theory postulates an 'economic man' who, in the course of being 'economic', is also 'rational'. This

---

* Allyn Young, a brilliant American economist who died relatively early through influenza, had illustrated the effects of the much more realistic assumption of increasing returns in an article in 1928 which was at least fifty years ahead of its time. Young had been head of the economics department at Stanford, had turned down the headship at Chicago, and held a chair at LSE when he died. So his work was very well known at the time – well known, but ignored.

man is assumed to have knowledge of the relevant aspects of his environment which, if not absolutely complete, is impressively clear and voluminous. He is assumed also to have a well-organized and stable system of preferences and a skill in computation that enables him to calculate, for the alternative courses of action available to him, which of these will permit him to reach the highest attainable point on his preference scale.

So far, all very relaxing and soothing to economists. But then comes his bombshell:

Recent developments in economics, and in particular in the theory of the business firm, have raised great doubts as to whether this schematized model of economic man provides a suitable foundation on which to erect a theory – whether it be a theory of how firms *do* behave, or how they 'should' rationally behave . . . [T]he task is to replace the global rationality of economic man with a kind of rational behavior which is compatible with the access to information and computational capacities that are actually possessed by organisms, including man, in the kinds of environments in which such organisms exist.

This latter quote essentially defines the research programmes carried out from the 1970s onwards by future Nobel Laureates such as Joe Stiglitz, Daniel Kahneman and Vernon Smith.

There are three distinct strands, to which each of these researchers is linked. First, the consequences for models based on the standard rational agent when either agents in general have incomplete information, or different agents or groups of agents have access to different amounts of information. With their predilection for grand phrases, economists refer to this latter as being a situation of 'asymmetric information'. Stiglitz has worked extensively in this area and is the source of much of its development. The other two strands blossomed into experimental economics and behavioural economics, inextricably linked to Smith and Kahneman, respectively. They overlap to a not inconsiderable degree and are often confused. Both relate to Simon's injunction to base theoretical models on agents which have 'computational capacities that are actually possessed by organisms'. In other words, to place realistic bounds on the ability of people to process the information which they have, regardless of whether it is complete. Again, the jargon phrase used by economists dresses this up as 'bounded rationality'.

The work on incomplete or asymmetrical information has been the easiest for mainstream economics to absorb, so much so that it is not only part of the standard toolkit of economists, but it has exercised powerful and pervasive influence on the conduct of economic policy. The seminal article, following Simon, and possibly the most brilliant single paper in the whole field of asymmetric information, is George Akerlof's 'Market for Lemons', published in the *Quarterly Journal of Economics* in 1970. 'Lemons' here does not refer to the fruit, but is used in the colloquial sense of a purchase that proves inferior to expectations. Akerlof illustrated his theoretical model with the example of the market in used cars, where the seller knows pretty well exactly how good the car actually is, whilst the buyer is less well informed. He deservedly shared the Nobel Prize in 2001 with two others: Michael Spence and Joe Stiglitz.

Stiglitz has become famous beyond the confines of economics for his espousal of what are often regarded as left-liberal positions on a range of important policy matters. The phrase 'left-liberal' does not imply that they are any more or any less correct, it simply describes the political views of many of the non-economists who follow his policy writings avidly. His distinguished work in economics rightly conveys status to his opinions. Fiscal deficits, globalisation, world poverty are all issues on which he has written extensively, and more recently he has supported the idea of the so-called 'Robin Hood' tax on every transaction carried out in financial markets.

Stiglitz is operating in the honourable tradition of political economy. The ultimate purpose of the study of economics is to gain a better understanding of the world, which might enable us to improve the human condition. It is desirable that its leading practitioners engage with policy debate rather than remaining cloistered in academia worrying about the role of the representative, rational agent in dynamic stochastic general equilibrium theory. I am not, incidentally, making this latter concept up, we will meet it in the next chapter. And we will see that this bizarre theory became very influential in central banks in the run-up to the financial crisis of the late 2000s.

Within economics, Stiglitz's major contribution has been to explore the implications of restricting the amount of information available to the rational agent of economic theory. To recap once more, this agent is postulated to have a fixed set of preferences, gather all information relevant to any particular choice, and then to make the best possible choice, the optimal choice, given his or her preferences. Again, all scientific theories are approximations to reality, and their usefulness depends strongly on how reasonable these approximations are.

The assumption that agents can gather all available information clearly restricts the empirical validity of the behavioural model of choice based on it. The realism of the rational economic agent is extended in its range once we allow imperfect information to be a part of the scientific model. Either when all agents have the same amount of imperfect information or, perhaps even more realistically, some have more than others, the so-called case of 'asymmetric information'. Imperfect information was one of the foundations which Simon laid for a new model of rationality. Stiglitz essentially erected the first proper building on these foundations, in the context of modern economic theory.

One brief but very important example will suffice to illustrate the concept, based essentially on the idea that there are two sets of agents: insiders, as we might term them, who have quite a lot of information in a given context, and outsiders, who have relatively little. The importance of this issue is shown by the fact that economists have a special jargon phrase to describe it, the so-called 'principal-agent problem'. The business firm based on the principle of limited liability of the shareholders has proved to be an enormously productive innovation. Companies have been and continue to be the key to the prosperity and the success of the developed economies. It is certainly possible to have large organisations able to benefit from economies of scale and scope in their operations which do not have shareholders – major accountancy and consultancy companies, for example, are often partnerships and not shareholder-based – but firms with shareholders are the dominant form of economic organisational structure in the Western economies.

Within these companies, there are two powerful groups, whose interests ought to work in synchrony but often do not. The shareholders own the organisation and they are the group which can exercise ultimate legal control over decisions. But, in outfits of any significant size, the owners – the shareholders – usually appoint professional managers to run them on a day-to-day basis. Again, there are exceptions, even in very large companies where the initial owner both continues to run the company and retains a substantial shareholding. Martin Sorrell of the multimedia conglomerate WPP is an example. Phillip Green of the UK retail giant Arcadia was another, although he transferred his shares to his wife based in Monaco, where in a single year she famously and entirely legally received a dividend of £1 billion.

The executives of a company are involved on a full-time basis with its operations and as a result acquire large amounts of information about its workings. Elaborate structures are put in place to ensure that such information is made available to shareholders as well, many of them with legal sanctions to back up any failures of disclosure. Yet, inevitably, it is simply not possible for the owners to be as well informed as the executive managers.

In the case of the aristocratic British bank Barings, the information asymmetry was dramatic. Nick Leeson, a trader based in Singapore, reported a consistent stream of large profits, from which the owners were only too pleased to draw their large dividends and the senior managers their bonuses. But it turned out eventually that the information was in general not only worthless – a fact very well known to Leeson, who was responsible for providing it – but to a large extent fraudulent, leading to the collapse and liquidation of the entire company. Bernie Madoff is an even more spectacular example of the asymmetrical distribution of information. Madoff knew that his scheme was based on a gigantic fraud. The unfortunate people who entrusted their money to him did not. This group includes not only individuals but professional fund managers such as Nicola Horlick of Bramdean Asset Management, based in Mayfair in the heart of London. Many clients were desperate to get their money into Madoff's funds, and it was regarded as a coup for Horlick when she was able to get such access.

More generally, in the run-up to the financial crisis of the late 2000s, there are numerous examples of cases where insiders – not in the narrow sense of illegal insider dealing but people who had a pretty good idea of what was really going on – were able to persuade outsiders, the shareholders, to adopt policies and follow strategies which turned out to serve their own short-term interests rather than those of the shareholders. To realise the scale of the problem we have only to recall a number of banks across the world such as Bear Stearns in which the shareholders lost all or almost all of their money but in which the executives were enriched. Most companies have a very distributed shareholder base, coordination amongst whom is very difficult to achieve and which therefore leads to a substantial weakening of the involvement of shareholders in company affairs.

Much of the current intense policy debate about appropriate structures and corporate governance of banks and other financial institutions centres on how best to handle this issue of the asymmetric information sets of executives and shareholders. There are other crucial issues as well, of course, but if their information could become more closely aligned, it is likely that their interests would converge as well.

*

The concept of asymmetrical information is a valuable one, but implicit in its workings is a related concept beloved of economists, namely that of 'market failure'. By this, the profession does not mean that markets fail in any way at all. Rather, the phrase describes situations where, for whatever reason, markets are unable to function as they would do if all the assumptions underlying the theory actually obtained: fully rational and fully informed agents and so on. A less snappy but more accurate alternative to the phrase 'market failure' would be along the lines of: 'in this particular context, there are factors which are preventing the market from operating as it should do'. Remember here Simon's phrase: 'how they [firms] "should" rationally behave'. The core model of mainstream economics does not purport to be merely descriptive – this is how the world works. It is prescriptive – this is how the world *ought* to work.

The theoretical world of general equilibrium is a seductive one. By a series of logical, mathematical steps, it can be demonstrated that such a world is perfectly efficient in the following sense. Prices are set so that supply and demand are exactly balanced in every single market. Everyone who wants to work has a job, for the supply of labour, the number people who want to work, is matched by an equal demand from employers. And there are neither unused resources nor shortages anywhere in the economy. It is perhaps not surprising that professional economists, after struggling in their graduate courses to master the fixed-point theorems needed to prove the existence of general equilibrium, often lose sight of the fact that this is merely a theory. And it is above all a theory of a static world, which tells us in principle how to achieve the best possible allocation of a given set of resources, of people, material, infrastructure and so forth. It does not tell us anything at all about how to create new, additional wealth and resources.

So economists have a description of the world which, if it obtained in reality, would possess a variety of apparently desirable characteristics. They slip rather easily into saying that this is how the world 'ought' to work. And if in practice that is not what is observed, there must be some 'market failure'. The world must be changed to make it conform to theory. So they devise schemes of various degrees of cleverness to eliminate such failures and enable the market to operate as it 'should', to allow rational agents to decide in a fully rational manner.

The concept of market failure has come to pervade policy making in the West over the past few decades, over a very wide range of policy questions. The role of the policy maker, in this vision of the world, is to ensure that conditions prevail which allow markets to work properly, and for equilibrium to be reached. Ironically, mainstream economics, with its idealisation of markets, now provides the intellectual underpinnings for government intervention in both social and economic issues on a vast scale.

In the UK, the Financial Services Authority is an example of a body created from exactly this mindset. Clever, rational people believed that other clever, rational people could devise written systems of rules, regulation and procedures which would ensure that situations could be created in which it would be as if markets were operating as they 'should'. In such circumstances, risks would be minimised, and possibly eliminated completely. The FSA was hailed at its launch in 1997 by Gordon Brown as 'a unique, twenty-first-century, one-stop centre, a single supervisor for all providers of financial services'. It has a staff of 2,500 people, is charged with enforcing no fewer than 8,500 pages of specific regulations,* and the direct costs of running it, excluding the costs to companies of complying with its myriad regulations, is now around £650 million (around $1 billion) a year.

As long as this rule book was followed by a financial company, the FSA was apparently satisfied. And despite its catastrophic failure to prevent the financial collapse of major banks such as RBS, this particular regulatory body in general continues to operate using exactly the same thinking: its purpose is to deal with market failure.

As an aside, illustrating the bureaucracy not in some Platonic ideal world where it is simply eliminating information asymmetries, but as it actually exists, readers will recall the terrifying month of September 2008 when the financial cataclysm burst. Lehman Brothers went bankrupt, and we seemed on the verge

---

* The new Conservative-led coalition government in the UK is committed to getting rid of the FSA, though whether this will change fundamentally the way in which any new regulatory bodies see the world remains to be seen.

of a repeat of the Great Depression of the 1930s. In October of that same year, barely a month after capitalism itself had returned from the brink of the abyss, a friend of mine at a hedge fund received a letter from the FSA. Its tone was stern. Indeed, its purpose was to issue a serious warning as to his future conduct. What had he done? Had he speculated irresponsibly to try to bring about the collapse of one of Britain's great banks? Had he sailed close to the wind and triggered concerns that he might be insider trading? Not at all. At the end of the month of September 2008, he had committed the grave offence of failing to send the Financial Services Authority the regular form in which he was required to document the activities he had carried out which had contributed to his professional development.

*

Simon raised the problem for the rational model of economic agents that they simply cannot gather all available information. The response from the profession has been to expand the theory to incorporate the concept, and to use it to justify massive interventions in the economy designed to eliminate 'market failure', to correct for the fact that agents may not have full information. But it is still fundamentally based on the idea that agents ought to behave as they do in a fully rational world. And, of course, these theoretical agents act entirely independently. Network effects are conspicuous by their complete absence.

Simon's second challenge was that, in many situations, agents lack the computational ability to work out what is the best choice to make, the optimal choice. Indeed, it may often not be possible to know what the optimal choice is, regardless of how we approach the decision.

The discipline of economics has responded to this in a way which is consistent with multiple personality disorder. At one level, and especially within macroeconomics, the problem raised by Simon for the economically rational agent has simply been ignored. It is 'as if', to use the favourite economic expression, the matter had never been raised. But at another level, within behavioural economics, some rather profound changes have taken place in how economists see decision makers, people, firms, governments making their choices. Agents are still presumed to be operating independently, but in ways which are markedly different from those of conventional economic rationality. As we shall see, there are some subtle differences between Simon's vision and the overall direction of this new research. But it represents progress nevertheless.

*

This aspect of Simon's vision has been taken forward by the two other Nobel Laureates, Vernon Smith and Daniel Kahneman, mentioned earlier in the chapter. As with asymmetric information, their work makes no use of the concept of networks and their impact on behaviour. But their empirical findings point the way to more realistic representations of how agents behave, albeit still as so many Robinson Crusoes, forming their opinions and making their decisions in splendid isolation.

Vernon Smith made the seminal contributions to the field of experimental economics. As the name implies, this is the study of economic situations using experimental methods, and during the course of which valuable data is gathered on how agents actually behave (as opposed to how they are posited to behave by mainstream theory).

Behavioural economics, linked irrevocably to Daniel Kahneman, is much more explicitly based upon experiments using methodologies developed in psychology to see how agents really do behave. And a key focus of such experiments is the actual rather than the theoretical cognitive capacity of agents. There is rather a fine line between this and experimental economics, and indeed in the rest of the book the distinctions will be ignored and the general phrase 'behavioural economics' will be used. Both provide valuable information on how people really do behave.

Like Herb Simon, Kahneman was not by training or profession an economist. Born in Israel in 1934, he obtained his degree in psychology from the Hebrew University of Jerusalem and he has made his academic career in the subject. He is currently attached to both the psychology and public and international affairs departments at Princeton, and in 2007 received the American Psychological Association's Award for Outstanding Lifetime Contributions to Psychology. Oh yes, and along the way he picked up the Nobel Prize in economics in 2002.

Kahneman's conclusions in his Nobel lecture are worth repeating here: 'The central characteristic of agents is not that they reason poorly, but that they often act intuitively. And the behaviour of these agents is not guided by what they are able to compute, but by what they happen to see at a given moment.'

*

There are many examples where people have been discovered to behave in ways which are pretty consistently different from the way the rational model of economics suggests that they should. The concept of 'framing' is one of them. This means that the choice a person makes can be heavily influenced by how it is presented. Volunteers in an experiment might be confronted with the following hypothetical situation and asked to choose between two alternatives. A disaster is unfolding, perhaps a stand is about to collapse in a soccer stadium and you have to decide how to handle it. Your experts tell you that 3,000 lives are at risk. If you take one course of action, you can save 1,000 people for certain, but the rest will definitely die. If you take the other, there is a chance that everyone will be saved. But it is risky, and your advisers tell you that it only has a one in three chance of working. If it doesn't, everyone will die. Simple arithmetic tells us that the expected loss of life in both choices is 2,000, for on the second option there is a two out of three chance that all 3,000 will be killed. When confronted with this, most people choose the first course of action.

The problem is then put in a different way, it is 'framed' differently. This time, you are told that the same first choice open to you will lead to 2,000 people being killed. The second will cause the deaths of 3,000 people with a chance of two out of three that this will happen, and one out of three that no one will die. The outcomes are identical to those set out above. Yet in this context, most people choose the second option.

Mainstream economists are often not persuaded by the results of experiments such as these. They are, after all, hypothetical situations and it is very unlikely that any of the participants would ever be called upon to make such a decision, or indeed anything remotely like it. There are many other aspects to this methodological debate, which is still ongoing between mainstreamers and experimental economists and which need not detain us.

For there are many real-life examples of the importance of framing. During the post-war period, trade unions in Britain gradually became more and more powerful, even bringing down Edward Heath's Tory government in early 1974 as a result of widespread industrial action. Heath's successor as Tory leader, Margaret Thatcher, was not only determined to change this when she came to office. She did. One of the ways by which she succeeded was by enabling an 'opt-in' rather than 'opt-out' method to be used as a way of deducting trade-union subscriptions from workers' wages. In some companies where trade unions were recognised by employers, everyone had to belong to the union, the so-called 'closed shop', and the company simply deducted the subscriptions and passed them on to the union. In others, membership wasn't compulsory, but it was assumed, and union dues were still automatically deducted unless the worker took the action of 'opting out', of filling in a form to say that he or she did not want to be in the union. 'Opting in' put the onus on workers to positively opt for union membership.

Almost exactly the same story is unfolding in Wisconsin and other American states as I write these words. Whether union dues are automatically deducted or whether workers have to opt voluntarily to pay and join is a very live political issue. The exact nuances vary from context to context, but the principle of framing is clearly at work here. How the choice is put can have a dramatic effect on the outcome, far greater than would be the case if it were a matter of 'rational' agents including the cost of the time spent filling in a simple form when assessing the various costs and benefits of trade-union membership. This might make a difference to a handful of people, those at the margin of joining or not. But in practice, the impact on the eventual outcome can be very strong.

Framing is just one of a number of concepts of behavioural economics which modify the standard way in which economists postulate that agents make 'rational' choices. In many ways, companies, and in particular their marketing departments, understood empirically the precepts of behavioural economics before academics got round to formalising the concept. They have always known that consumers often behave differently from the way that standard rational agent economic theory would have them behave.

Marketeers observed, for example, that discount offers such as 'buy one, get one free' or 'three for the price of two' – a concept

I am very keen on because this is how bookstores often package up their offers – tend to be more effective in boosting sales than the exact equivalent price reduction on a single purchase. The amount of money which is paid for the bundle of products is identical in each case, but more will usually be bought if they are packaged under an offer than if there is a simple equivalent reduction in the individual prices.

Marketing departments also understand that consumers can be very impatient and will pay to get something sooner rather than later. Turning to online shopping, the excellent range offered on Amazon makes the site extremely popular. The company offers an option to deliver for free within three to five working days. Alternatively, customers can pay to have their item delivered within two days, and even more to have it delivered the next day. Now, there will be circumstances in which the latter option is desirable, such as forgetting it is your wife's birthday until the very last minute. But many items – books, for one – are bought to be savoured, to last, and they may very well be re-read over a long period of time. So paying a premium, often a non-trivial percentage of its price, to have it the next day rather than just a few days later does not make 'rational' sense. But it is an option nevertheless routinely chosen by many consumers. Again, in practice it is often found that the introduction of a considerably more expensive option into a range often results in consumers switching from cheaper options to what is now the second most expensive brand. Few people will buy the newly introduced expensive product, but what was previously the most expensive is now perceived as somehow being better value than it was. The prices of all the previously available items remain unchanged, but people will switch away from cheap ones to the now second-most expensive.

These are all examples of how agents often behave differently from how the standard rational choice model would have them make choices, and give a flavour of the work of behavioural economists, which by now is voluminous.

*

Kahneman himself generously describes a 1980 article by the University of Chicago scholar Richard Thaler as the 'founding text in behavioural economics'. Thaler indeed is deservedly a major figure in behavioural economics. In 2009 he published a best-seller with Cass Sunstein entitled *Nudge: Improving Decisions About Health, Wealth and Happiness.* The dialogue with the authors on the amazon.com page for the book defines 'nudge' very clearly: 'By a nudge we mean anything that influences our choices'.

There are many examples, and here are just two. Two Yale academics, Dean Karlan and Ian Ayres, set up stickK.com, a website designed to help people reach their goals. So a person who wants to lose weight, say, makes a public commitment on the site to lose at least a specific amount by a specific date. He or she agrees a test to be carried out, such as being weighed in the presence of named witnesses, to verify whether or not the goal has been reached. But the person also puts up a sum of money, which is returned if the goal is met. If not, the money goes to charity. The second example is the 'dollar a day' plan in Greensboro, North Carolina, aimed at reducing further pregnancies in teenage girls under sixteen who have already had a baby. In addition to counselling and support, the girls in the pilot scheme were paid a dollar for each day in which they did not become pregnant again. Of the sixty-five girls in the scheme, only ten of them got pregnant again over the next five years. Of course, there are many criticisms of these and other such 'nudge' concepts. A persistent and strong one is that the people who really do want to lose weight are the ones who make the commitment, the girls who really do not want to get pregnant again are the ones who join the scheme. In other words, those who sign up to 'nudge' schemes are those who were likely to adopt this behaviour regardless. Nevertheless, 'nudge' remains an influential concept.

'Anything which influences our choices' is a definition which clearly includes the concept of incentives in the familiar guise of a monetary cost or benefit. But it also includes less obvious ones, such as making a public commitment to reach a goal. And we can even think of examples from the opening chapter as being examples of 'nudge', such as the 'price' of being burned to death or firing a shot in the air to signal a group of potential hooligans to disperse. But it is broader than that, as the example of framing suggests. It is not simply a question of incentives, but how they are presented.

These are all ways of extending the concept of incentives, which may or may not work in any given context, but which are nevertheless a useful intellectual construct. Less helpfully, it is claimed that 'nudge' appears to be able to solve almost any problem, certainly as far as policy makers of almost any description are concerned, as another exchange in the amazon.com interview makes clear.

Amazon: 'What are some of the situations where nudges can make a difference?'

Thaler and Sunstein: 'Well, to name just a few: better investments for everyone, more savings for retirement, less obesity, more charitable giving, a cleaner planet, and an improved educational system. We could easily make people both wealthier and healthier by devising friendlier choice environments, or architectures.'

The scene from the Monty Python film *Life of Brian* springs irresistibly to mind, when a member of the revolutionary People's Front of Judea haranguing a crowd asks rhetorically, 'What have the Romans ever done for us?' only to be provided in reply with a list including clean water, sanitation, roads, wine, education, peace. 'But apart from that?' he asks plaintively.

Incentives matter. Nudge widens the concept of how incentives operate, and Thaler has made a distinguished contribution to our scientific understanding. But might these claims be a little overstated? More wealth, more health, nicer people, a better environment, all can be achieved 'easily'. This has not so far, it should be said, been the experience of the Nudge team set up at the instigation of the current British Prime Minister, David Cameron, and advised by Richard Thaler himself, although to be fair it has only been in existence since 2010.

*

Behavioural economics advances our knowledge of how agents behave in practice. But in many ways it sidesteps the most fundamental challenge which Simon posed to rational economic behaviour. He argued that in many circumstances, we simply cannot compute the 'optimal' choice, or decide what constitutes the 'best' strategy. This is the case even if – especially if! – we have access to complete information in any particular situation. In many situations it is not just that the search for the optimal decision might be time consuming and expensive, it is that the optimal decision *cannot* be known, at least in the current state of knowledge and technology.

This is an absolutely fundamental challenge to the economic concept of rationality. Many economists nowadays choose to interpret Simon's work in ways which are compatible with their theories. Sure, they argue, there are often limits to the amount of information which people gather, the amount of time and effort they take in making a decision. But this is because they judge that the additional benefits which might be gained by being able to make an even better choice by gathering more information, spending more time contemplating, are offset by the costs of such activities. Agents may well use a restricted information set, and make an optimal decision on this basis. An even better choice might be made if more information were gathered, but not one sufficiently better to justify the additional time and effort required.

But this whole argument completely misses Simon's point. He believed that, in many real-life situations, the optimal choice can *never* be known. It is not just a question of being willing to spend more time and effort to acquire the information so that the truly optimal choice can be made. We simply cannot discover it.

A tragic, but nevertheless sadly rather routine, recent story from the north of England encapsulates this point. A ten-year-old boy drowned trying to save his step-sister who had fallen into a deep pond. Two police community officers who were present refused to enter the water, on the grounds that they had not been trained in water safety. Incredibly, at the inquest a detective chief inspector defended this behaviour. Given their lack of training it would, he explained sanctimoniously, have been 'inappropriate' for them to try to save the child's life.

But this was not the reaction of the human beings, in contrast to the tick-box robots, present at the scene. The brave young boy instinctively tried to save his sister. Two fishermen, both well into their sixties, leapt into the pond without thinking or training. They rescued the girl, but the boy died.

We can readily generalise from this single incident (although the true rational planner would dismiss it as 'anecdotal evidence'). Ex ante, there is usually no single best course of action to follow. Jumping into a pond of unknown depth risks your own life and in any case you may be too late. Standing by and doing nothing means the girl will die, but you will live. The future is fraught with both risk and uncertainty, inherent parts of the human condition, and therefore of our social and economic systems.

Incidents such as these occur on an almost everyday basis. But although there are always many self-righteous individuals willing to pontificate after the event on what 'appropriate' behaviour should have been, this does not mean that the 'optimal' behaviour can be identified even with the benefit of hindsight. The agents involved had to make a decision quickly, but they still had fairly complete information. Local knowledge meant that all the agents knew that the pond was deep and cold. The girl could not swim. What is the optimal course of action for anyone witnessing this unfolding tragedy? Even in a relatively simple situation such as this, there is too much chance and contingency involved to be confident of any analysis. Too many tiny details, whose interactions could lead to success or failure, for us to say what the best strategy of any of the agents involved would have been. The fishermen saved the girl, but, especially given their ages, the shock of plunging into the cold water may have induced cramp or even a mild heart attack, leading to the death of one or even both the brave rescuers. We simply cannot say.

In his 1955 paper, one of the illustrations which Simon used to make his point was the much more innocuous pastime of chess. There are very good reasons for his choice. In chess, agents do indeed have complete information. Not just a good selection of all the information which potentially exists, but literally all of it. The rules of the game are not only rather straightforward and few in number (around a dozen), but they are also fixed. So both players have total understanding of the rules. The game can be won in only one way, by capturing the opponent's king. The number and strength of each player's remaining pieces is irrelevant if one of them is able to checkmate. Indeed, many of the most brilliant games between grandmasters are based upon material sacrifice to achieve precisely that aim. So the purpose of the game is clear. And both players have complete information about what the other side has already done, the action is transparent.

But Simon points out that in the typical game, after fifteen to twenty moves or so on each side, a sequence of the next sixteen moves, eight by each side, might be expected to yield a stupendous total of permissible variations. He calculates the typical number of such variations in such positions as being approximately ten followed by twenty-four zeros. And this is just for the next eight moves on either side; often from these so-called 'middle-game' positions, in practice there will a further twenty moves each before the game is concluded, and sometimes many more. Clearly, it is literally impossible for any human, even aided by a computer, to work out so many variations. Left to our own devices, even at an average of one per second it would take longer than the lifetime of the universe.

Simon considered how players respond to such situations. Rather, he considered the much more general point of what the principle of rationality might look like in situations when the complications exceed the capacity of humans to compute all eventual outcomes. He simply used chess as a particular illustration of his thinking. And the game of chess is simplicity itself when we consider many of the decisions we all have to make on a more or less everyday basis. Choosing a partner, selecting a pension scheme, deciding whether to take another job, casting a vote in an important election, all these are clearly very complicated with potential consequences which last far into the future. Yet they occur with reasonable frequency. Even apparently much more simple situations are difficult to analyse in advance, as we saw above with the tragic incident of the boy who drowned in the pool.

When driving back to London, where I have lived for many years, from visiting my father in my home town I might equally well take one of three motorway routes. As I approach the motorway I face an immediate choice: turn left or turn right. Both are valid choices, involving completely different routes of approximately equal distance. Whichever I pick, on the approach to the vast London conurbation the number of potential routes proliferates rapidly. In common with most motorists, I use various satellite navigation aids to provide information. These, indeed, confirm the huge number of ways in which the journey can be made, well over 100,000 routes being examined by the machine when I set off. My choices through the journey are influenced by traffic information, but even this is not straightforward to interpret. I have to form a view on how long the delays have been in place, whether they will have cleared by the time I reach them and so on. And, of course, new delays may emerge after I have made any particular choice.

So how do I choose? Although I have made the journey many times, it has not been and never will be possible for me to sample more than a tiny fraction of the potential route combinations, and even if I did, the next time I chose a particular one the circumstances might have changed.

I use what as an economist I would describe as 'heuristics', but writing in English I call instead 'rules of thumb'. These are basic guidelines which an agent evolves, drawing on his or her experience over time, both particular to the choice in question and also possibly more widely. On any particular occasion, a given rule might not work in the sense that it leads to an undesirable outcome. I set off, and two hours later find myself in a major traffic jam owing to a serious accident which has only just happened. So the rules are not fixed, but might be modified over time.

A key feature of such rules is that they do enable me to scale down the problem of choice dramatically, to enable me to consider realistically no more than a dozen, or at the very most twenty, of the actual six-figure number of alternatives I face.

The overall trip is around 220 miles. But the shortest in terms of distance cuts around fifteen miles off this figure. It is a route I never use. Why? Because it involves crossing almost the whole of Greater Manchester, an urban area with a population of some 2.5 million. I know from my general experience that travel on such roads tends to be slow and that motorways are faster. There have been a number of memorable occasions over the years when it would almost certainly have been better to take this alternative, but these are few and far between and so I use this rule of thumb to discard not just this, but also other non-motorway routes over the course of almost the entire journey, until London itself is entered.

In terms of the first choice, left or right at the first motorway junction, encountered after a mere handful of miles, my rule of thumb has changed over time. Satellite navigation aids are not really helpful here, they indicate a right turn but in the absence of known traffic delays the left-turn journey is estimated to be only three minutes longer. I used to turn left and now turn right. This is because I began to encounter longer delays simply due to heavy traffic in the last thirty miles or so through London and its approaches. Just heavy traffic plain and simple, nothing which would show up as a delay on a Satnav system. It might now be quicker on average to use this route, but I don't. The right-turn route, despite problems, in general has given satisfactory outcomes.

And here is another essential feature of rules of thumb. Agents use them as long as they continue to deliver not optimal, but satisfactory outcomes. Indeed, even once I have arrived safely at home after completing one of these trips, I simply do not know which of the alternatives might have been the best, the 'optimal'. What I do know is whether the journey I have just experienced has been satisfactory.

These are exactly the processes which Simon described chess players using in his 1955 paper. They use rules of thumb to scale down dramatically the dimension, the scale of the number of possible alternatives to evaluate. And provided a rule of thumb continues to give reasonable results, they continue to use it. They realise that the search for the 'optimal' decision is very often completely pointless, and so settle for a viable decision which is likely to be satisfactory.

*

In principle, I suppose, the best route on the motorway could be calculated once the journey had been completed, but not before. Tracking by satellite of all the potential routes as I made my way back to London, and subsequent analysis of the data might be able to reveal what on this occasion would have been the optimal route. But this is a time-consuming process requiring very advanced technology. In many apparently simple situations, even the most advanced technology cannot compute the best decision. Think of the tragedy in the pond described above, and then try to think of the details of the information which would be required in order to decide what would have been the best course of action for everyone concerned.

This is why, again, Simon used the example of chess to illustrate the limits to knowledge. Chess is simple to describe, and players have complete information. But in a large proportion of all the stupendous number of potential positions, the optimal cannot be known, even using advanced technology in the form of computers. These are now very distinctly stronger than the best human players. Even the world champion now has no chance of winning if matched in a series of games against a computer.

In chess, even a beginner knows that the queen is by far the most powerful piece on the board in terms of its attacking powers, and the pawn the weakest. So sequences of moves which involve you obtaining an opponent's pawn at the cost of your own queen can be immediately discarded (unless, of course, there would be a consequential benefit such as checkmating your opponent's king). In most situations, there are many such possible sequences, less dramatic than this but which involve material loss with no apparent compensation.

But the use of rules of thumb goes much deeper than this in the game of chess. There are thirty-two pieces on the board at the start of the game. Computers have now solved completely all possible positions in the game – when there are only six pieces left on it! Some of these are far beyond the capacity of even the strongest human to compute, requiring sequences of over 200 moves, and at each stage in these cases literally the optimal move has to be selected. To put this in context, no more than a handful of games have ever been played between strong players which have 200 moves in total, and complete games over even 120 or 130 moves are rare. Some progress has been made by computers in solving positions with seven pieces, but the task is far from complete, even using very powerful machines.

In fact, between reasonably strong players operating at their best, differences in their abilities to calculate sequences of possible moves in any given position ought not to be that decisive in determining the outcome. Of course, in practice under competitive conditions, stress, nerves, or tiredness may often lead to mistakes, and some very strong players may indeed be able to 'see' a few moves further ahead in the sense that they can calculate all relevant sequences, and this can sometimes be a factor. But players at high levels differ more markedly in their ability to form a judgement on the likely result which will follow from a future sequence of moves, once the ability to calculate further moves through precise analysis is exhausted. Both players may 'see' a particular position after the next, say, ten possible moves perfectly. One may judge it a win for White, say, whilst the other may judge it a draw. Neither can possibly calculate such an outcome; both are relying on their experience and judgement to arrive at this view. So both sides readily obtain the resulting position, and the one with the superior judgement usually prevails (unless a significant error is made in the course of the rest of the game).

Indeed, in many positions in a game, there is little point in carrying out elaborate calculations on possible sequences of moves. Many famous positions, some of them decisive in world championship matches, have been extensively analysed by the very strongest players for decades after they were actually played. The results are published, available to everyone to scrutinise and to improve. But the best, the 'optimal' move in the crucial position remains unknown. Even Gary Kasparov, probably the strongest human player ever, equipped with modern computers, has been unable to solve some of these problems in his monumental five-volume series of books on all previous world champions, entitled *My Great Predecessors*. Incomprehensible to someone who has never played chess, they are nevertheless replete with practical illustrations of Herb Simon's concept of rationality.

Playing chess obsesses a small number of people, and interests a much larger number, but if chess disappeared tomorrow most of humanity would not notice. Selecting a route for a driving journey is a very humdrum, everyday task. Men of certain ages and predilections may enjoy fiddling with the technology, though for most people it is a tedious, routine activity. But, as we have seen, even in what we might describe as humble situations, far, far removed from complicated life-or-death decisions, it is often not possible to calculate the optimal choice. Agents use rules of thumb both to scale down the dimension of the problem and to select amongst alternatives.

It is always possible for true believers in the economic concept of rationality to describe any conceivable outcome, any conceivable apparent mode of behaviour, as being consistent with the postulates of the theory. It is, as they say, 'as if' I were choosing the optimal decision, given the various probabilities involved, in deciding my route, or 'as if' the world champion chess player were making the optimal move. The Communist Party of the Soviet Union made use of a marvellous phrase: 'it is apparent'. 'It is apparent that . . .', the Party would pronounce, before going on to assert something which was at best a half-truth and was often a downright lie. So, in this spirit, let us say that it is apparent that if an agent failed to take the optimal decision, he or she would eventually learn to take it.

*

We met circumstances in Chapter 1 – the behaviour of the mob, the courage of the martyrs – where ex post a story could always be told which would reconcile the outcomes and behaviour with economic rationality. But both these and the more everyday examples used here stretch the credibility of this way of thinking to breaking point. And especially when, applying Occam's razor, there is a simpler explanation available, which fits more comfortably with the empirical evidence. Agents often have imperfect information but, much more fundamentally, lack the computational capacity to calculate the optimal decision, even with the limited amount of information they do possess.

To emphasise again, this is really quite a different concept of rationality from the one defined by the postulates of conventional economics. The idea that there is only one definition of rationality, that of mainstream economic theory, has become firmly embedded in policy discourse. But it is completely wrong. If the assumptions, the simplifications, required by standard theory are reasonably congruent with any given empirical situation, then, yes, it is rational for agents to behave in this way. If they are not, we need a new definition of rationality, one which recognises not just the fact that not everyone has all relevant information at any point in time, but, more fundamentally, that many situations are so complex that, in the current state of scientific knowledge, the optimal decision cannot be computed by any conceivable agent. At some indeterminate point in the future, when they are incredibly more powerful than they are today, computers will eventually solve the game of chess completely. And maybe then we will be part of the way towards enabling agents to compute optimal decisions in everyday situations which, as we have seen, are often considerably more complicated than chess.

In other words, we need a definition of rationality which is consistent with the social and economic worlds of the twenty-first century, which accounts for the nature and limitations of human cognition, one which completes Simon's vision. Economics dismisses behaviour which does not conform to its precepts of what is rational as 'irrational'. But on the contrary, it is economic rationality which is itself irrational in many modern contexts.

We may note as an aside that very recent scientific developments go even further. Simon's concept of rationality is anathema to the believers in economically rational optimising behaviour. Fortunately for the blood pressure of the latter, such people are in general rather narrow minded and rarely read scientific material outside their own immediate field of interest. An article in the top journal *Science* in July 2010 by Dutch psychologists Ruud Custers and Henk Aarts would have burst more than a few blood vessels. The title – 'The Unconscious Will: How the Pursuit of Goals Operates Outside of Conscious Awareness' – gives a pretty good guide to its content.* The opening paragraph states boldly:

* This is often not the case at all. My old Cambridge tutor Christopher Bliss deservedly made his academic economics reputation with a paper entitled 'On Putty Clay'.

As humans, we generally have the feeling that we decide what we want and what we do. These self-reflections remind us that we are not bound to the present environment for our actions. We can envision ourselves in different places, in alternative futures, doing different things. We only have to decide to do so, and we can go and see a movie tonight or hang out with friends in a bar. It is up to us. Our behaviors seem to originate in our conscious decisions to pursue desired outcomes, or goals.

It goes on: 'Scientific research suggests otherwise', and concludes that 'the basic processes necessary for goal pursuit – preparing and directing instrumental actions and assessing the reward value of the goal – can operate outside conscious awareness'. The psychologists do, however, concede that 'it is too early to conclude that consciousness is redundant in the pursuit of goals'. But perhaps this line of thought is a little too radical even for the most iconoclastic economist. So let us proceed on more familiar ground.

The concept of limits to human computational abilities was not discovered by Simon. Indeed, we can trace it back at least as far as Aristotle, who wrote in the fourth century BC: 'It is the mark of the instructed mind to rest satisfied with the degree of precision to which the nature of the subject admits and not to seek exactness when only an approximation to the truth is possible.' It may not be possible to discover the best strategy, the optimal choice, because 'only an approximation to the truth is possible'. Not: 'only an approximation to the truth is used', as the modern economic distortions of Simon's key message would have us understand, that the best could be obtained but we just do not think it is worth it. But: 'only an approximation to the truth *is possible*'.

*

The same idea was expressed in modern economic terms by Keynes. His magnum opus, the *General Theory of Employment, Interest and Money*, was published in 1936, and immediately became the focus of intense theoretical controversy. It is the work for which Keynes is remembered. Much less well known is his 1937 article in the *Quarterly Journal of Economics* in which he set out to clarify aspects of his book.

Here is Keynes, anticipating Simon by nearly two decades: 'we have, as a rule, only the vaguest idea of any but the most direct consequences of our acts', he says. Only the vaguest idea! But Keynes was a practical man as much as a theorist, and he goes on: 'the necessity for action and for decision compels us as practical men to do our best to overlook this awkward fact and to behave exactly as we should if we had behind us a good calculation of a series of prospective advantages and disadvantages'.

He asks rhetorically: 'How do we manage in such circumstances to behave in a manner which saves our faces as rational economic men?' In other words, Keynes here opens the possibility of rational modes of behaviour which are quite different from those of the mainstream, equilibrium economic model.

And he answers his own question, describing what Simon would later call 'rules of thumb'. It is worth quoting this at some length:

We have devised for the purpose a variety of techniques, of which much the most important are the three following:

1   We assume that the present is a much more serviceable guide to the future than a candid examination of past experience would show it to have been hitherto. In other

words, we largely ignore the prospect of future changes about the actual character of which we know nothing.

2 We assume that the existing state of opinion as expressed in prices and the character of existing output is based on a correct summing up of future prospects, so that we can accept it as such unless and until something new and relevant comes into the picture.

3 Knowing that our individual judgement is worthless, we endeavour to fall back on the judgement of the rest of the world which is perhaps better informed. That is, we endeavour to conform with the behaviour of the majority or average. The psychology of a society of individuals each of whom is endeavouring to copy the others leads to what we may strictly call a conventional judgement.

So when agents have only the vaguest idea of the consequences of their decisions, when by implication they can never discover with any reasonable degree of accuracy what they might be, they adopt heuristics which scale down the dimension of the problem.

In his points 1 and 2, Keynes is essentially arguing that we give much greater weight to the current situation than is warranted when forming a view about the future. Certainly, the experience of the past few years supports this idea. During the financial and economic boom of the late 2000s, most people – regulators, governments, international bodies such as the International Monetary Fund (IMF) – thought it would go on for ever. When the bubble burst, everyone became plunged in gloom. The Nobel Prize-winning economist Robert Barro stated in the spring of 2009 that the prices prevailing on the various financial markets were signalling a 30 per cent chance of a repeat of the Great Depression of the 1930s. We might think the recent experience has been bad, but during it output (GDP) fell by only 3 per cent

in America and by 5 to 6 per cent in the major European economies. In the 1930s, output fell by nearly 30 – thirty! – per cent
in both Germany and the US. That is why it is called the Great
Depression. And markets were indicating a 30 per cent chance of
a repetition of that experience, which fortunately has not taken
place.

It is Keynes's third point which is the most explosive, and
which anticipates by many decades modern network theory, how
agents are connected to each other, and how, where and when
an agent bases its actions at least in part on the behaviour, opinions or decisions of others. In short, Keynes argues that copying the behaviour of others is in fact rational – the 'conventional
judgement'.

Agents can rarely, if ever, know in advance with any reliable
degree of accuracy the consequences of their actions. They have
only the vaguest idea. This opinion fits in very well with Simon's
radical and innovative concept of rationality. But it is still one
whose explicit base is the autonomous agent, making decisions as
an independent entity.

But Keynes gives a foretaste of the effect of introducing network effects into the principles of agent choice. It is in fact is a
natural extension of Simon's principle of rationality. And at the
same time it carries such radically different properties, such different implications, that it should be regarded as a separate definition in its own right.

Keynes argues that a good 'rule of thumb' on which to rely is
to copy other people. In other words, to base your decisions (at
least in part) on what other people do, other people to which you
are connected and of whose actions and decision you are aware.
In short, your network.

Effects, as we have seen, can spread across networks both for
good and for ill. In the run up to the financial crisis, for example,

the belief spread across networks of traders, networks of regulators, networks of politicians, that economics had finally solved our problems and the boom would go on forever. But it ended, as we shall see in the next chapter, with waves of doubt and pessimism percolating across a range of different networks. The trick for successful policy, for positive linking, is not which interest rate to try to manipulate, not whether to increase taxes or cut spending. It is the subtle but elusive goal of enabling the right frame of mind to spread across the networks which connect the relevant decision makers.

Herb Simon's programme for revolutionising the way we think about how people behave, how decisions are made, went deeper than just giving us a better way of thinking about how incentives operate, valuable though this may be. He argued that in many situations agents lack the ability to process the information, that our computational capacities are insufficient to deal with the complexities of the real world.

*

Most of the discussion in the book so far has been about what economists would call 'micro' issues. Some, such as the choice of a route to get from A to B or the playing of chess, have indeed been rather humdrum, everyday activities. Others, such as criminal activity, are of more substance and importance for society as a whole. Still others, such as the choice of religious faith or political ideology, have implications not just for society as a whole, but in the contexts in which we discussed them could literally be a matter of life and death for the relevant individuals.

We have already seen examples of where network effects can overwhelm the impact of incentives, however sophisticated or 'nudged' they may be. When people are influenced by networks, they are not just operating in a completely different way from that posited by rational economic theory, using rules of thumb rather than trying to optimise. People are operating in contexts where the capacities of even the most sophisticated policy maker to compute with a high degree of confidence the consequences of a decision are not just stretched to their utmost, but are inadequate for the task.

Of course, in practice, a mixture of incentives, whether conventional ones or in their more modern 'nudge' guise, and network effects will operate. But once network effects start to influence the outcome in any meaningful way then, as we have seen, the difficulties for the policy maker appear to increase considerably.

But we have not yet really touched on what economists call 'macro' issues, namely the behaviour of the economy at an overall, aggregate level. Keynes has now been mentioned, primarily for his insights into how people actually behave, how they act in a way which he describes as rational but which is utterly different from the rationality of the economics textbooks.

Keynes is mainly remembered as a macroeconomist, remembered for his great work the *General Theory*, in which he contemplated how it could be that the system of liberal capitalism, which he greatly admired, could have brought about a situation in which the total output of economies fell by nearly a third, and in which nearly one in four Americans were unemployed.

So in the next chapter, we look at some macroeconomic issues, and specifically the experience of the recent financial crisis. How did it come about? What was the role of rational agent economic theory? And what role was played by networks, of agents making decisions on the basis of how others behaved? The role played by what Keynes called 'the psychology of a society of individuals each of whom is endeavouring to copy the others'.

4

Did Economists Go Mad?

The conduct of economic policy making over the ten to fifteen years prior to the financial crisis of 2008–9 exemplifies the fundamental problems of the conventional mindset of economics. At the time, it seemed as though clever policy makers devising clever rules and regulations to set the right incentives, to which economically rational agents would respond appropriately, had indeed solved key problems of macroeconomic management. Economic growth in the West was strong and steady, and both unemployment and inflation everywhere remained low.

Networks were conspicuous by their complete absence from the intellectual framework of policy makers. Yet network effects were absolutely central to the causes of the crisis.

There had been a number of scares along the way. In 1997–8, the rapidly growing area of East Asia experienced a financial crisis, with huge falls in output and employment throughout the region in 1998. But very soon growth and rising prosperity were restored. Network effects featured strongly in both the crash and the recovery. The countries of the area were held up as shining examples of success, with rapidly rising prosperity and high levels of investment in education and infrastructure. Doubts began to emerge about the economy of Thailand. In particular, there were worries – harbinger of bigger things to come ten years later – about whether the country was experiencing an unsustainable real-estate bubble. The Thai currency came under attack, and the government cut its link, its fixed value, to the US dollar. This was the signal for massive speculation against the currency and a sharp fall in its value.

But the crisis then spread like wildfire across almost every single country in the region. Even China, then nowhere near as connected to the rest of the world as it is now, suffered from a loss of foreign confidence. The financial collapse led to dramatic falls in output in many of the countries, with soaring unemployment. In Indonesia, the thirty-year rule of the dictator Suharto collapsed after widespread rioting in protest. So events which were particular to Thailand cascaded across the networks which shaped opinion in financial markets, and the entire region came under speculative attack. In US dollar terms, the output of countries such as Thailand, Indonesia, Malaysia and South Korea fell by more in a single year than the worst affected economies in the Great Depression of the 1930s.

Then, almost as suddenly, confidence returned, across both the networks of financial markets and the networks which create business confidence, or lack of it, in the domestic economies of the region. Why? Well, as the old saying goes, success has many fathers and failure is an orphan. The IMF was widely blamed for its role during the crisis, but claimed credit for restoring stability and paving the way for recovery. Even now, despite over a decade of the most intensive study by economists, we do not have an agreed answer to the questions why the economies collapsed so spectacularly but then, very quickly, bounced back as though nothing had happened. Networks have played little part in any of this analysis, but were undoubtedly the key.

One of the most dramatic knock-on effects of the crisis in East Asia was on an economy outside the region: Russia. There were genuine reasons to be concerned about the financial health of the Russian economy. Taxes were not being collected, many state employees were not being paid for months at a time, and the government had already been forced to raise interest rates to an astronomical 150 per cent to try to stop even more money from fleeing the country. The East Asia crisis sparked even further doubts, if that were possible, about the Russian situation. In the late summer of 1998 the Russian government defaulted on its domestic debt and stopped paying interest on its foreign debt.

At around the same time, in the United States the collapse of Long-Term Capital Management, a hedge fund with two economics Nobel Prize winners in its luminaries, sparked a temporary panic. LTCM lost nearly $5 billion in less than four months and was forced to shut down completely in 2000. But the panic was temporary. The problem appeared solved and the world moved on. Exactly the same thing happened in the aftermath of the bursting of the dot.com bubble at the start of the new millennium.

So, a massive crisis in East Asia, a default on its debt by the Russian government, a huge failure of a speculative fund in America, the collapse of the dot.com bubble. Any single one of these might have triggered a worldwide crisis. But what might at any moment have turned into a bloodbath seemed to have been averted by the new-found skill and knowledge of policy makers, equipped not just with the longstanding tools of incentives but with the insights provided by the concept of 'market failure'.

The intellectual underpinnings for the apparent miracle were provided by economic theory. Olivier Blanchard is the chief economist of the IMF. Here is what he had to say in August 2008 in an MIT working paper entitled 'The State of Macro': 'For a long while after the explosion of macroeconomics in the 1970s, the field looked like a battlefield. Over time, however, largely because facts do not go away, a largely shared vision both of fluctuations and of methodology has emerged . . . The state of macro is good.' The state of macro is good! In August 2008!

A few weeks later Lehmans went bankrupt. Capitalism itself was on the brink of another Great Depression on the scale of the 1930s when unemployment in the USA reached nearly 25 per cent. The period which had been dubbed the Great Moderation, when policy makers seemed to have been granted the touch of King Midas,* in reality proved to be the Great Delusion.

*

One of Keynes's most well-known phrases refers to the power of ideas. In his *General Theory* he wrote, 'practical men, who believe themselves to be quite exempt from any intellectual influences, are usually the slave of some defunct economist. Madmen in authority who hear voices in the air are distilling their frenzy from some academic scribbler of years back.'

But in contrast to Keynes's view of the role of ideas in the crisis of the 1930s, the crisis of the late 2000s was grounded not in ideas which were advanced by academics 'years back'. It arose from ideas which play a prominent role in contemporary academic economics. Far from being 'defunct', these ideas became more and more important in the decade or so leading up to the crash in 2008.

There were two distinct strands to the intellectual mindset which believed that the right structures to correct market failure, the right incentives, the right appreciation of asymmetrical information, could solve all problems.

---

* Except of course the hapless British finance minister Gordon Brown who 'modernised' the UK's holdings of gold and foreign currencies by selling all the country's gold reserves when the price was at a near-record low. Why hold gold when all the uncertainties and problems of economic management have been solved? That was the thinking. At the time of writing, this decision has cost the British taxpayer at least $20 billion.

The first goes back about forty years, to a trio of American academics leading successful but blameless careers. The practical significance of their seemingly esoteric work has been immense. The intellectual impact of their findings was just as important, if by no means as obvious. But this, too, had huge practical significance in the policy stance which governments and regulators took towards financial markets. Fischer Black, described by one of his close friends as 'the strangest man I ever met', soon left academia to make millions at Goldman Sachs before his tragically early death. Robert C. Merton and Myron Scholes received the Nobel Prize in economics in 1997 for their findings.

These three discovered ways of applying concepts from statistical physics to financial markets. These concepts enabled prices – incentives – to be established in a vast number of markets which had scarcely existed before.

Their findings enabled the creation of today's industry of financial derivatives, worth over $500 *trillion*, according to the Bank for International Settlements. The basic idea of derivatives – so called because their value is derived from, or related to, that of an underlying asset – is very simple. Suppose an investor holds some Vodafone shares. He or she may worry that the price will fall. Someone else may think it will rise. A contract can be struck between them to trade the shares at a specified price at a date in the future. Its price will vary depending upon the price of Vodafone shares at any point between now and then.

The crucial feature of derivatives is that their price tends to fluctuate much more than that of the underlying share to which they are linked. The rewards of getting it right can be much bigger, but so too can the losses. Vodafone has traded for some time in the region of 150p a share, so suppose for illustration that this is the price today. If I feel optimistic about the company, I can buy the shares now. If in a month's time, say, they are 300p, I will have doubled my money. But I could instead buy *the right* to buy them at, say, 250p in a month from now for virtually nothing, 1p perhaps, since it is so unlikely that such a big increase will happen in such a short time. If I am proved right and the price really is 300p, I have the right to buy shares at 250p and can then sell them immediately for 300p. My 1p has turned into 50p, far, far more than doubling my money. But if I am wrong and the price stays below 250p, I will lose everything I put in.

In short, derivatives both satisfy and create an appetite for risk. They enable much riskier bets – sorry, considered investment judgements – to be made than if you can just trade in the underlying shares themselves.

As it happens, Merton and Scholes got their comeuppance when they totally misjudged some risks and their financial company, LTCM, mentioned above, collapsed in 1998 with a loss of nearly $5 billion, and had to be bailed out by the Federal Reserve. So today's problems are not exactly without precedent.

But Black, Merton and Scholes had initiated a period of stupendous innovation in financial markets. The introduction of incredibly high-powered mathematics into financial markets created all sorts of hitherto undreamt-of possibilities.

*

The major intellectual challenge in economic theory from the late nineteenth century for almost a hundred years was to specify as precisely as possible the conditions under which the operation of the free market could be guaranteed to be efficient, in the sense of establishing the theoretical existence of an equilibrium in which the price mechanism – incentives – would ensure that all markets in the economy cleared. That supply would equal demand in every market, and so no resources would be left unused. The details of this need not concern us here. Suffice that it is an exceptionally difficult intellectual problem, and no fewer than seven out of the first eleven Nobel Prize winners in economics received their award in whole or in part for their work on this problem. As we have seen, many economists have come to believe that this is a description of how the world *ought* to behave, and incentives and regulatory structures should be put in place to achieve this aim.

By the late 1960s, the problem – the existence of a so-called 'general equilibrium' – was essentially solved completely, with many of the necessary results having been established in the 1950s. However, there was a rather embarrassing implication which even economists of a high theoretical bent could not fail to notice. Essentially, for the existence of general equilibrium to be established, an assumption had to be made that there were complete markets.

'Complete markets'. Surely a rather obscure and innocuous phrase? But it means that a market must in principle exist for any transaction at any time. A price must in principle be able to be determined. So, to take two illustrative examples almost at random, a price has to be able to be set today for a purchase of sterling with dollars on 23 May 2041, and also for the purchase of the 1904 edition of Wright's *Grammar of the Gothic Language* on 15 September 2028. With a bit of scouting around, it may be possible to strike a deal on the former. But the latter? It is not even possible in practice to set a price for the weekly purchase of groceries in a month from now. All these are markets, and all require a price to be set.

The transparent lack of most such markets regarding future transactions led Kenneth Arrow, perhaps the single most distinguished contributor to the proof of existence of general equilibrium, to describe it in 1994 as being an 'empirical refutation' of the theory.

But here is where the academic trio rode to the rescue. One of the practical issues was that economists did not really know how prices should be established in this plethora of markets extending into the indefinite future which was required by the fundamental theory of free-market economics. Black, Merton and Scholes appeared to solve this problem. Their formula seemed to be able to determine what the price 'should' be for any transaction in any market in the future. Complete markets might not be observed in practice, but economists now seemed to know how to set prices in any market.

One of the products which could now, apparently, be priced 'optimally' was nothing other than risk itself. The markets would ensure that the right incentive, the right price, was in place to capture accurately the risk on any particular transaction, no matter how complicated.

One of these goes by the name 'securitisation'. The concept itself has been around for a long time. What was different in the run-up to 2008 was the use of risk pricing to cut up the securitised package into different packages and then create markets in which the packages could be bought and sold. It is this obscure and seemingly anodyne concept which made a major contribution to the subsequent financial crisis. A bank makes loans to a large number of individuals, the normal practice of banking since time immemorial. It collects fees for making the loans, and then, in the usual course of events, receives the interest due on them.

The innovation of securitisation involved bundling the loans up into a package, and selling the package on to a separate company, created by the bank itself. The risk was therefore taken off the bank, which still kept its fees. The separate company found the money to buy the package by issuing securities. These securities were bought by sophisticated financial market operators, who could then in turn sell them to someone else, almost like shares on the stock market. It was this latter operation, the crea-

tion of markets in securitised products, which was the real innovation. The value at any point in time would depend upon how
the individual loans were performing and on how different people assessed the risks involved in buying the package. But this was
now something which could safely be left to the market; everyone involved had the incentives to ensure that the risk was priced
accurately.

The first real harbinger of the financial crisis occurred in the
autumn of 2007, just a year before the real catastrophe took place.
It led to the collapse of the British bank Northern Rock, a relatively small, geographically concentrated bank with a good reputation and an apparently sound business model. Yet its demise led
to the sight on television of panicked savers queuing to try to get
their monies out, the first time this had been seen in the UK since
the nineteenth century.

The maths of pricing many of the individual parcel-passing
trades is so hard that even the theoretical physicists doing it
couldn't always be relied upon to get it right. It is certainly far
beyond the capabilities of most board members in even the most
august financial institutions.

This last point really gets to the nub of this first phase of the
crisis. A system had been created which was so complicated,
so convoluted, that even at the very highest levels in financial
companies, no one really understood the level of risk which was
being carried at any point in time. We might recognise the sur-
roundings. They are none other than Herb Simon's strictures on
the limits to human computational ability, and hence cognition.
Literally no one had understood the full ramifications of the
world which had been created. Gradually, doubts began to seep
across the network of banks, doubts about whether it was possible to know the true potential extent of losses of another bank
to which money had been lent. Once banks became uncertain

about whether they understood the true financial positions of other banks, they became reluctant to lend to each other.

Indeed, in August 2007 they simply stopped doing so, more or less completely. A real, no-holds-barred credit crunch.

The problem was not specific to any one bank, not specific to any incentives, any specific pricing of risk which had been undertaken. It was a network effect. A network effect which gripped most of the world's banking system. One moment everything seemed fine and banks were happy to buy and sell these very complicated securitised bundles of loans. The next, almost in a twinkling of an eye, they were not. Indeed, they were extremely reluctant to carry out almost any sort of inter-bank trade, and specifically they stopped being willing to lend.

The price of each individual, isolated transaction had apparently been set optimally, the risk associated with it had been correctly assessed and taken into account. But banks had to believe that this was so. The belief had to be sustained across the networks on which the opinions and sentiments of bankers are formed. In Keynes's phrase, the bankers were 'a society of individuals each of whom is endeavouring to copy the others'. If they copied the opinion, the belief, that everything was fine, it would continue to be so.

But once they stopped believing, we had a credit crunch.

*

This was a decisive illustration in the run-up to the September 2008 crash of the robust yet fragile nature of networks which we encountered in the opening chapter. Banks, regulators, governments believed that the problem of pricing risk had been solved. Despite occasional doubts, occasional shocks, this belief persisted. Then, suddenly and dramatically, doubts about this spread like the Black Death across the networks of sentiment that run through financial institutions. And once this had happened, unlike Peter Pan exhorting the children to believe in fairies to save Tinkerbell from death, the authorities – regulators, governments, international institutions – found it impossible to exhort the banks to believe. Networks swamped all their efforts to restore confidence. Pessimism spread like wildfire.

To be fair, even though the authorities were not looking at the situation from a network perspective at all, there had been several prominent examples in the previous two decades of the robust nature of networks. Economies *can* withstand shocks and emerge relatively unscathed. Examples such as the East Asian crisis were mentioned at the start of this chapter. But there had been even more.

One Monday in October 1987, for instance, completely out of the blue, stock markets collapsed. In a single day, the value of the world's biggest companies fell by 20 per cent. But the sky didn't fall. Pessimism did not spread. Indeed, in Britain, the excesses of the boom of the late 1980s continued unchecked. House prices continued to soar, City traders continued to quaff champagne and collapse in stupefaction on their trains home. It was only two years later, for entirely unconnected reasons, that this particular party came to an end.

Even more telling is the experience of Japan over the past twenty years or so. By the late 1980s, Japan was *the* success story of the post-war era. Once derided for their cheap and nasty unreliable products, Japanese companies had come to bestride the globe. Visiting American bankers were obliged to overcome their squeamishness and consume live lobster sashimi in deference to their hosts.

Yet in 1990 pessimism suddenly infested the economy, and during the year the Nikkei share index lost 40 per cent of its value, bottoming out in 1993 at 80 per cent lower than its peak of just under 40,000. Even now it is only around 9,000, less than one-quarter its level of twenty years ago.

The collapse in land values was even more complete. In the late 1980s, rumours abounded that individual golf courses in Tokyo were worth more than the entire real estate of the state of California. But prices fell in large parts of the market by no less than 90 per cent.

Imagine. You are living in a house apparently worth half a million. You wake up the next day and find its value slashed to £50,000. Surely this would precipitate mass pessimism and a recession just as bad as the American one of the 1930s?

Logic says it would. But it didn't. In 2008 Britons and Americans were each, on average, about 40 per better off compared to the late 1980s, whereas in Japan the increase was just under 20 per cent. Not brilliant, but very far from being a disaster. Quite how the Japanese avoided catastrophe is still a bit of a mystery. Certainly, not just once but several times most of the major Japanese banks became technically bankrupt. With great aplomb, the Japanese central bank simply changed the rules and said they no longer were.

But the dramatic contrast between America and Japan shows that mass psychology, the percolation of pessimism or optimism across business and consumer networks, is almost impossible to judge in advance. In America in 1930, shocks in financial markets led to a stupendous collapse of the economy. In Japan in 1990, in apparently similar circumstances, things just bumbled along. Somehow, the authorities in Japan pulled off a very difficult feat of positive linking. They convinced everyone, financial markets, Japanese firms and consumers, that the spectacular collapse in land and equity prices were not about to plunge Japan into a repeat of the Great Depression of the 1930s.

Governments, central banks and international institutions like the IMF were bolstered in their Panglossian view of the world by intellectual developments within macroeconomics. We have seen above how the apparent ability to form complete markets and price the risk of each individual transaction 'optimally' proved seductive. Further enticement was provided by the concept of dynamic stochastic general equilibrium, or DSGE for short. Not perhaps the first phrase which sprang to the mind of the great seducer Lothario when in pursuit of one of his conquests, but one which proved just too beguiling for the authorities to resist.

The concept of the economically rational agent still has a firm grip on mainstream economics. But much of the exciting research in microeconomics, the study of individual agent behaviour, over the past three decades or so has distanced itself from this paradigm. As we have seen, Herb Simon's manifesto has by no means been carried out in full and has been the subject of attempts to neuter its most fundamental and radical message. But many economists now have a more realistic, more empirically grounded view of individual agent behaviour.

Incredibly, in macroeconomics the intellectual trend has been the complete opposite. The rational agent has emerged as the central character, the *only* character, in accounts of how the economy behaves at the overall, macro level. Imagine a *Macbeth* in which Shakespeare had fused Macbeth, his wife, Banquo and Duncan into a single character, and had written the play with no other characters at all. Imagine it? 'Rather tricky,' we Brits might say in British English, meaning 'almost impossible'. But this is exactly the sort of bizarre world in which modern macroeconomic theory invites us to believe.

Ideas such as these at the heart of modern macroeconomics have provided the intellectual justification of the economic policies of the past ten to fifteen years. And it is these ideas which the recent crisis has shown to be false. The dominant paradigm in macroeconomic theory over the past thirty years has been that of rational agents making optimal decisions under the assumption that they form their expectations about the future rationally – the rational agent using rational expectations.

Rational expectations do not require that an agent's predictions about the future are always correct. Indeed, such predictions may turn out to be incorrect in every single period, but still be rational. The requirement is that on average, over a long period of time, expectations are correct. Agents are assumed to take into account all relevant information, and to make predictions which are on average unbiased. Deviations from perfect foresight in any given period are an inherent feature of this behavioural postulate, but such deviations can only be random. If there were any systematic pattern to the deviations, the agent would be assumed to incorporate the pattern into his or her expectations. Again, on average over a long period, such expectations would be correct.

It will be apparent that the theory is difficult to falsify to someone who really believes in its validity. Even the most dramatic failure to predict the future, such as the 2008 financial crisis, can be explained away as a random error. A rational expectations enthusiast can continue to maintain the correctness of the theory by simply assuming that on average, over some (theoretically indeterminate) period of time, agents' expectations prove accurate.

An assumption of the theory is that, as part of the set of information being processed, the agent is in possession of *the* correct model of the economy. Indeed, on the logic of the theory itself, if the model being used to make predictions were not correct, the forecasts would exhibit some sort of bias, some systematic error, and agents would realise that it was wrong.

It might reasonably be argued that it is difficult to subscribe to the view that agents can even recognise the correct model of the economy, given that economists themselves differ in their views as to how the economy operates. In the autumn of 2008, many prominent American economists, including a number of Nobel Prize winners, vigorously opposed any form of bail-out of the financial system, arguing that it was better to let banks fail. Others, including decision makers at the Federal Reserve and Treasury, took a different view entirely.

The response of the academic mainstream has been to insist that there have been strong moves towards convergence within the profession on opinions about macroeconomic theory. By implication, anyone who takes a different view and is not part of this intellectual convergence is not really a proper economist.

A – possibly *the* – major project in macroeconomics over the past thirty-odd years has been to try to use equilibrium theory and the rational agent, rational expectations view of the world to explain the dynamic fluctuations in output which have been observed in the developed, market-oriented economies ever since the Industrial Revolution.

There are two aspects of this. First, the slow but steady growth in output over time, averaging around 3 per cent a year. It is this long-run growth which distinguishes capitalism from all other forms of social and economic organisation in human history. (This is a major topic in its own right, of course; I am merely mentioning it in passing here.)

The second is the persistent short-term fluctuations in output around this underlying slow growth. From time to time, these fluctuations are severe and output actually falls for a period of time, before growth is resumed. We in the West have just lived through one of these periods and, as I write, some people continue to believe another might well be imminent.

*

A theory based upon equilibrium appears to have an inherent problem when confronted with data such as that in Figure 4.1.

This shows annual percentage changes in total output in America from 1900 to 2010. It is entirely typical of the Western economies.

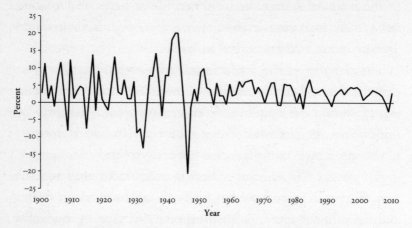

Figure 4.1 Annual percentage change in real US GDP, 1900–2010

Understanding why these fluctuations take place is very difficult. If they were understood in the same way as, say, building bridges is understood, we would have a pretty complete grasp as to why most of the world's developed economies are in their current predicament. But we do not. Given that these fluctuations are persistent both over time and across countries, they represent a serious challenge to a *Weltanschauung*, the framework through which the world is interpreted, based on the concept of equilibrium.

The first major attempt was 'real business cycle' (RBC) theory, developed in the 1980s by Finn Kydland and Edward Prescott. RBC has been very influential in mainstream economics, its originators receiving the Nobel Prize in 2004. According to this theory, periods of high or low growth – the booms and busts of everyday parlance – are initiated by random shocks to the economy. There are many problems with this theory, not least of which is the identification of what these shocks actually are, but the most widely used shock in RBC models is that of random changes in productivity.

In such models, recessions arise because of the rational response of individuals to adverse productivity shocks.* In a further illustration of the rather Orwellian use of words by mainstream economics, the 'real' of RBC signifies that recessions are caused by 'real' factors such as productivity and rational behaviour by agents. 'Real' is juxtaposed with 'nominal', nominal factors being such obviously irrelevant concepts as money, credit and debt!

Agents maximise utility over time, choosing between consumption and leisure. They have two decisions to make in every period. First, how much of their time to spend at work producing output (income) and how much to take in leisure. Second, how much of this output to allocate to investment, which will increase future levels of output, and how much to consume now.

Focusing just on the first of these may illustrate why even many fairly mainstream economists failed to be persuaded by the RBC approach. A temporary reduction in productivity today encourages people to work less now than in the future, because they will earn relatively more per hour in the future than they do today. So they choose to work less now. Some may work sufficiently less for it to seem as if they are unemployed, whereas according to RBC, they are actually rational agents maximising their expected lifetime utility by choosing to minimise their working hours. US Economist Paul Krugman famously noted that this account of the world suggests that the Great Depression, when nearly one in every four workers in America was unemployed, was essentially an extended voluntary holiday.

I have used the word 'agents' in the above description. But this is inaccurate. Strictly speaking, I should use it only in the

* All attempts to identify monetary factors as the source of shocks in this theory have failed.

singular, not the plural. For in these models, and their development into DSGE, as already mentioned above, there is only one agent. In the jargon, this is known as the 'representative agent'. I am not making this up. We are seriously invited to believe that the complicated workings of huge economies can be understood at the macro level by reference to a model containing only one agent, deemed to represent everybody, every firm, every government, every international body, every regulator. It is the complete antithesis of networks. They do not matter at all, because there is only one agent. Only one character in the play. Unless the agent started to trade with its counterpart on Mars – a hypothesis no less plausible than the concept of the representative agent itself – there is literally no one else with whom to form connections.

That aside, a rather glaring defect of the approach is that in financial crises, the responses of creditors and debtors may very well be, in fact almost invariably is, quite different. So just as a single-actor *Macbeth* is incomplete, as a very minimum any reasonable model needs two categories of agent: debtors and creditors.

Dynamic stochastic general equilibrium models are, just like RBC theory, based on the key microeconomic assumptions of orthodox economic theory. In other words, rational utility maximisation by consumers, rational value maximisation by firms, both operating under the assumption that they form expectations about the future rationally.

Essentially, DSGE models build on the RBC framework by trying to incorporate some features of the real world. So in RBC models, prices are very flexible and adjust rapidly to prevailing economic conditions. Under DSGE postulates, firms exercise some degree of market power and so prices may be more 'sticky' and take time to adjust to their new equilibrium levels following a shock to the system.

Although they are very complicated and difficult to construct, these models rapidly became very influential in academic economics.

These developments were not mere ivory tower musings. They rapidly gained influence with policy makers. Not of course that finance ministers themselves were intimately familiar with the nuances of DSGE models. But their advisers certainly were. As Olivier Blanchard of the IMF wrote just prior to the crash: 'DSGE models have become ubiquitous. Dozens of teams of researchers are involved in their construction. Nearly every central bank has one, or wants to have one. They are used to evaluate policy rules, to do conditional forecasting, or even sometimes to do actual forecasting.'

Note in particular the sentence 'Nearly every central bank has one, or wants to have one.' DSGE models became the central banker's fashion accessory of the moment – everyone else has one, so I must, too! Even within the rarefied portals of the world's central banks, network effects were present in terms of the fashion for copying such models.

So when politicians proclaimed the end to boom and bust, they had enormously powerful intellectual authority behind them, the models of the major central banks, the leading orthodox academic economists and the leading economic journals. They really did believe they had solved the macro problems of the Western world.

In the brave new world of DSGE, the possibility of a systemic collapse, of a cascade of defaults across networks connecting agents in the system, was never envisaged at all. Indeed, it is simply not possible in most such models, consisting as they do of just a single 'representative' agent. The models by their very nature ruled out the feature which distinguishes almost all financial crises in the history of the world: the loss of confidence which spreads rapidly across networks for bankers, traders, speculators.

*

Despite these apparent major intellectual advances, it should be said, forecasters continued to make exactly the same mistakes which they used to make when I started off as a macro modeller and forecaster way back in the 1970s. The problem was general amongst forecasters. Figure 4.2 is a chart from the October 2008 *Bank of England Quarterly Bulletin* about the revisions made to forecasts for GDP growth in 2009 as we progressed through 2008.

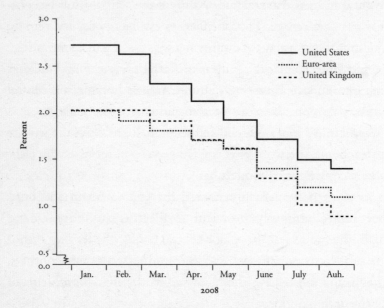

Figure 4.2 Predictions of real GDP growth for 2009 made during 2008. Source: Bank of England Quarterly Bulletin, October 2008

So at the start of 2008, decent growth was predicted for 2009. Even as late as August, the general view was that there would still be positive growth in 2009. But in fact, the West was already in recession in August 2008 and growth was already below zero! In actual fact, output fell by nearly 3 per cent in the US in 2009, by

5 per cent in the UK and by 6 per cent in the eurozone.

This was not simply a one-off error in an otherwise exemplary forecasting record. The major crisis in East Asia in the late 1990s was, as we have noted, completely unforeseen. Here are some figures to back up the assertion. In May 1997, the IMF predicted a continuation of the enormous growth rates which those economies had experienced for a number of years: 7 per cent growth was projected for Thailand in 1998, 7.5 per cent for Indonesia and 8 per cent for Malaysia. By October, these had been revised down to 3.5, 6 and 6.5 per cent, respectively. But by December the IMF was forecasting only 3 per cent growth for Malaysia and Indonesia, and zero for Thailand. Yet the actual outturns for 1998 for these countries were spectacularly worse, with output not growing but falling by large amounts. The fall in real GDP in 1998 was 10 per cent in Thailand, and 7 and 13 per cent in Malaysia and Indonesia, respectively.

So what happened next when the crisis struck? How was the world saved?

In the week of 15 September 2008 capitalism nearly ground to a halt. Share prices collapsed. Credit markets froze. And we were within hours of cash machines, ATMs, being closed to the public.

It was the American authorities who really saved the world in that terrifying week. And they did so not by the manipulation of elegant rational expectations models and theories, but by experiment and by relying on their knowledge of what had gone wrong in the Great Depression of the 1930s. Faced with a wholly uncertain immediate future, the authorities reacted by trying rules of thumb, by seeing what worked and what did not. They reacted exactly as Herb Simon said humans behaved all those years ago. They knew it was impossible to work out the optimal strategy. So they tried things which seemed reasonable and, quite literally, hoped for the best.

It was fortuitous – and an important illustration of the role of chance and contingency in human affairs – that the chairman of the Federal Reserve at the time, Ben Bernanke, was a leading academic authority on the Great Depression. He knew that, above all, the banks had to be protected. It may seem monstrously unfair that the bankers themselves escaped penalties – indeed, it *is* unfair – but the abiding lesson of the 1930s is that in a financial crisis the banks have to be defended. Money is the blood which flows through the economy to keep it alive. If the chairman instead had been, say, a world expert on dynamic stochastic general equilibrium models, we would almost certainly now be in the throes of the second Great Depression.

Bernanke had already restored a concept which is absent from the rational behaviour rule book, that of 'moral suasion'. Moral suasion, the central bank 'persuading' bankers to make particular decisions, is how central banks used to operate before the complicated, rule-based, hugely expensive bureaucratic control systems based on concepts of 'market failure' were introduced. Bear Stearns was a massive global investment bank which got into serious financial difficulties through massive losses on dealings in sub-prime mortgages. In March 2008, the company received an emergency loan from the Federal Reserve Bank of New York. It became apparent very quickly that this would not be sufficient to save the bank.

Using moral suasion, Ben Bernanke persuaded another bank, J. P. Morgan to take it over, with all its potential liabilities, in the course of a weekend. J. P. Morgan was under no legal or rule-based obligation to do so. But, somehow, they were persuaded, and also persuaded to pay $10 a share instead of the initial agreement to buy at just $2. (This, incidentally, is the same method used by the Bank of England to solve the previous banking crisis in the UK way back in the early 1970s.) The shareholders of Bear

Stearns still lost out dramatically, the peak share value over the previous twelve months having been over $130, but an immediate major financial collapse was avoided.

Of course, six months later the whole banking system was teetering on the edge. But why then and not previously? Why then and not even at all? Debt. That is the word that everyone began to worry about, and it is still an important source of concern in 2011. But why? Figure 4.3 charts debt in America from 1920 until immediately before the financial crisis in the autumn of 2008. Naturally, over such a long period the population grows, the economy grows, prices change. We cannot simply compare the amount of debt out there in mere money terms. What the chart does is divide the total amount of outstanding debt owed by people and companies, including financial ones, by the size of the economy.

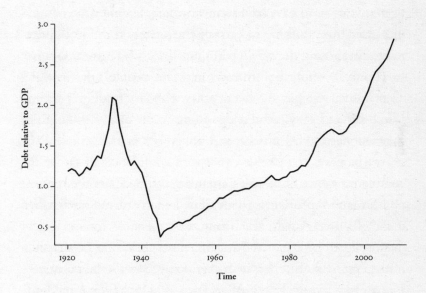

Figure 4.3 Total private debt in America compared to the size of the economy. Source: Steve Keen, University of Western Sydney

Shock, horror! Compared to the size of the economy, debt in 2008 was even higher than it was in 1929, the blip in the left-hand side of the chart, just before the Great Depression. The value of debt owed by individuals and companies is now nearly three times the size of the economy. Surely this meant that an economic collapse was inevitable?

The striking feature is the continuous rise in debt compared to the size of the economy over the entire post-war period. But the post-war period has been a time not of economic gloom but of entirely unprecedented rises in living standards. A willingness to take on debt can often be a sign of confidence about the future. For individuals, a confidence that things will get better and the debt be repaid.

Even more importantly, companies with new plans, new ideas, need loans in order to translate them into reality. These often fail, but when they succeed, the rewards can be spectacular. The last twenty years or so have seen some now famous American companies grow from nothing to become the biggest in the world, dramatically altering the way we live our lives – Microsoft, Google, Facebook to name but three. There are people now working away in their garages in Silicon Valley with visions of overturning Microsoft – if they are actually going to do this at some point they will need to incur debt, and probably lots of it.

Again, it was the elusive concept of confidence, or rather the complete absence of it, across financial and business networks of opinion and sentiment, which caused the crash to occur when it did. In every single year from 2000 onwards, private sector debt as a percentage of the American economy was higher than it had ever been. But despite scares along the way, the party continued, just like in Edgar Allan Poe's *Masque of the Red Death*. Then, suddenly, sentiment in financial markets became universally pessimistic.

Admittedly, the authorities did try the experiment of allowing Lehman's to fail. But it rapidly became evident that such a laissez-faire policy risked the collapse of the entire Western capitalist system. No monetary authority since has seen fit to repeat the experiment.

Much publicity and controversy surrounded the setting up of the resulting Troubled Asset Relief Program (TARP), a $700-billion bail-out fund which required political approval and so was played out in full light of the democratic process in America. But in many ways this was of second-order importance to the purely administrative actions of the American authorities, who:

* nationalised the main mortgage companies, Fannie Mae and Freddie Mac;
* effectively nationalised the gigantic insurance company AIG;
* eliminated investment banks;
* forced mergers of giant retail banks; and
* guaranteed money-market funds.

This last in particular has attracted very little attention, but was probably the single most important specific measure which was taken. The money-market funds hold very short-term assets, and are consequently obliged to hold highly liquid, high-quality assets. Indeed, the funds are essentially required to hold a dollar of assets for each dollar lent. But on 16 September, the Reserve Primary Fund* wrote off Lehman Brothers' stock, and the value of its shares fell below the critical dollar mark, to 97 cents. This almost triggered a massive run on the banking system as a whole. If this had happened, an immediate consequence would have been that ATM machines would have been closed and consumers would have had difficulties getting hold of cash. It is rumoured

---

* This was its name; it had no connection with the US Federal Reserve.

that the relevant sub-committee of the British Cabinet met to consider the risks of major public disorder if this actually happened. Companies would not have been able to roll over their short-term debt, and if they did not have cash in hand to cover what they owed, they would have had to file for bankruptcy.

In short, the default of money-market funds could easily have triggered by itself a massive recession. But on 19 September, the US Treasury announced that it would guarantee the holdings of any public money-market fund which participated (for a fee) in the programme.

The key point about all these actions is that the American authorities paid no attention to academic macroeconomic theory of the past thirty years. RBC theory, DSGE models, rational expectations – all the myriad erudite papers on these topics might just as well have never been written. Instead, the authorities acted. They acted imperfectly, in conditions of huge uncertainty, drawing on the lessons of the 1930s and hoping that the mistakes of that period could be avoided. It was not a grand plan, nor did one ever exist. This was a process of people responding to events on the basis of imperfect knowledge and experimenting to discover what did and did not seem to work, desperately trying to restore confidence across financial networks. And networks were important to the outcomes, to the decisions which were made, at a very detailed level.

Confidence across networks was key. Confidence across networks of financial institutions that the monies owed to them by others would be paid. Confidence across networks of commercial companies that output was not about to collapse like it did in the 1930s, so that they would then not act in ways which made this a self-fulfilling prophecy. And confidence across networks of individuals that their worlds were not about to fall apart.

*

The accounts of the seemingly perpetual meetings which took place between bankers, regulators and policy makers at the time make it clear that personal relationships, the interactions between the major players as human beings, also played a key role. If different people with different personalities had been involved, even if they too had thrown academic macro theory out of the window, the outcomes would have been different.

What they came up with worked. American GDP in 2009 fell by some 3 per cent compared to 2008, and by the autumn of 2011 the economy had not only stabilised but had grown for nine successive quarters. Indeed, by the third quarter of 2011, growth was sufficiently strong that the level of US GDP rose above its pre-crisis peak. (In contrast, between 1929 and 1930, the first year of the Great Depression, GDP fell by nearly 9 per cent, and the cumulative drop between 1929 and 1933 was 27 per cent.) Unemployment was still high, but employment had risen. The stabilisation programme worked, and a catastrophic collapse in output during 2009 was averted. It prevented pessimism and panic from percolating across networks.

It is a spectacular success of positive linking. The specific details of the measures which were taken were in general of second-order importance compared to the success in preventing the sentiment from spreading that America was about to suffer a re-run of the Great Depression of the 1930s.

The crisis in Europe during 2011 has mainly arisen through a failure of eurozone governments to generate anywhere near the same degree of positive linking of sentiment across financial networks. In late 2011, as the final revisions are being made to the book, it is still unclear as to what the eventual outcome will be. We can usefully think of much of the economic policy in Europe during 2011 as being not about specific measures in particular, the sort of thing which economists get excited about and whose impact they continue to believe they can measure, but about the desperate attempts by key policy makers to spread positive sentiment across the markets.

One thing is clear. Confidence is only weakly related to objective reality, to the actual facts. The principal concern is about public sector debt. In the case of Greece, the concern is entirely merited. Compared to the size of output in Greece (GDP), public sector debt is above 150 per cent, and there are few encouraging signs of a willingness to get to grips with the problem. With interest rates at around 7 per cent, this means that some 10 per cent (7 per cent of 150 per cent) of the total spending GDP Greece is going not on providing any form of services, but on paying the interest on its debt.

The comparable figure for Spain is only 60 per cent. Yet Spain, too, has experienced repeated crises of confidence in the markets, and its interest rates have hovered around 7 per cent. In contrast, the interest rates on government debt in both the UK and Germany are not much more than 2 per cent, even though public sector debt is 80 per cent of GDP in the UK and nearly 85 per cent in Germany. Obviously public debt is not the single cause of the lack of confidence, but the Japanese currency is perceived of as being strong despite a public debt ratio of nearly 200 per cent.

Policies which generate confidence are, once again, not so much the specific details, the economically 'rational' calculation of their potential consequences, but the creation of a positive mind set, a positive attitude on financial markets. Positive linking.

Such momentous events fully deserve the economists' description of 'macro', a prefix which is a straight transliteration of the Ancient Greek μακρo, meaning 'big'. Incentives were at work, in addition to network effects, not least because policy makers and regulators believed that they had put structures and incentives in place which would ensure that risk had been priced optimally and, in consequence, had been neutered. But, eventually, this belief was totally shattered by the impact of networks.

The world of the economically rational agent, forming expectations rationally, was both an intellectual and a practical mirage. The real world proved to be completely different. A world characterised by 'the psychology of a society of individuals each of whom is endeavouring to copy the others'. A world in which the optimal decision can never be known, where decision makers of all kinds fall back on Simon's rules of thumb. And a world in which the unexpected happens all the time.

But how far do these principles extend? As we have seen, they are certainly relevant not only to the economy at the macro level, but also to 'macro' issues which impact on many people, such as crime or ideological and religious beliefs. But do they only apply at the 'macro' level, using the word in its simple, original sense? We need to spend some time in the much more trivial, 'micro' world of popular culture to gauge how pervasive these factors might be. And the opening examples in the next chapter fit the description 'trivial' to perfection.

5

Lady Luck: The Goddess of Fortune

Ben Nevis is the highest mountain in the British Isles. Situated in the Highland region of Scotland, by world standards it is a mere pimple, standing at just 4,406 feet (1,344 metres). But even by the tourist path and the good track which it provides, hill walkers feel every single inch in the mere four miles it takes to get to the summit. And despite its modest height, the summit can be a brutal place. The mountain experiences the direct blast of the prevailing winds from the Atlantic, and gusts of 100mph are not uncommon. The mean average temperature is less than one degree above freezing. Snow can fall there on any day of the year, I have actually witnessed such a fall in August, at the height of so-called summer, from the comparative safety of a slightly lower hill where the precipitation merely fell as sleet.

The M1 in Britain is the main route out of London leading to the Midlands and the north of England. It was the first inter-city motorway to be completed in the UK, the main sections being built between 1959 and 1968, and carries a vast amount of traffic. Unusually, the traffic density reduces somewhat in the approaches to inner London, after the junction with London's orbital motorway, the M25, this extension having been opened in 1977. In April 2011, this part of the road was closed to all traffic for a few days because of a fire which had taken place immediately under one of the elevated sections.

These two seemingly disparate paragraphs, replete with details of wholly trivial facts, are connected. Videos of people doing their ironing on both sites were posted on YouTube in April 2011. One man struggled to the summit of Ben Nevis with an ironing board as well as a rucksack strapped to his back, the iron itself presumably being in the sack. Another man, unshaven and wearing a dressing gown, ironed a shirt over a period of three minutes on the closed lanes of the M1.

One of the videos received over 4,000 times as many hits on YouTube as the other. As the man in the more popular clip was reported as saying in a subsequent interview in a Sunday newspaper: 'I feel sorry for [him] . . . putting the video on YouTube only to receive forty-six hits. I climbed [. . .] and now my video has 204,000 hits. My story has been covered in Japan, America, Greece and Russia. People probably think I'm a mad Englishman, but I don't care.'

The missing words in the quote following 'I climbed' are 'through a gap in a fence and walked for two minutes to the motorway'. So it was the M1 ironer who experienced worldwide interest in his feat, if we might use this word to describe his doings, and the intrepid climber who disappeared into the mists of time.

The difference in the degrees of interest shown in these two events is very hard to explain in the framework of the isolated, rational agent reacting to incentives. As usual, we could in principle always tell a story after the event which purports to account for the much greater popularity of one video compared to another. But it is even harder to make up in this context than it was with the incident of the rioting soccer fans in Sardinia in the opening chapter. At least with the hooligans, the view of the world in which autonomous agents respond to incentives could offer a reasonable explanation for part of what happened, even though by no means the whole. Here, the imagination quails at even trying to offer a rationalisation in the incentives frame-work. Eccentric Englishmen iron clothes in unusual situations. The intrinsic similarities in the two events are profound. Yet one proves 4,000 times more popular than the other.

This is by no means an isolated example. On the contrary. In the world of popular culture, using this term in a broad sense, such massive discrepancies of outcome between virtually identi-cal 'products' is quite literally an everyday occurrence. Most lan-guish in obscurity, whilst a select few become very popular, for no obvious reason other than sheer good luck.

The most popular daily download on, say, Flickr or YouTube will usually be not just several thousand but several hundred thousand times more popular than most of the photos or videos which are uploaded on any given day. Occasionally, there will be an intrinsic, obvious reason for the most viewed or downloaded choices, such as footage of a tsunami taken by a lucky survivor. But most of the time, this is not the case.

I am writing this paragraph on Easter Monday. On the Flickr website today I see that 'Hot tags' in the last twenty-four hours include 'Easter', 'Easter morning', and 'resurrección'. Not at all surprising. But there is also 'rund', a tag of German origin con-taining photos of wholly disparate objects connected only by the quality of roundness, and 'Animal Collective'. The latter is not some modern communal or Soviet-style experiment in animal husbandry, but the word 'experiment' can indeed be used to describe what it actually refers to, for Wikipedia assures me that Animal Collective is an experimental rock band from Baltimore currently based in New York City. Another hot tag is 'specialty', which, on inspection, after the first few shots of various cakes, appears to consist of hundreds of photographs of an entirely ordi-nary dog doing entirely ordinary things.

The phenomenon of a lack of connection between any inherent, objective quality of an offer and its popularity relative to other, very similar things is not confined to leisure activities such as viewing and downloading from websites. It is increasingly widespread across markets where people pay real money, even for what we might in a Neanderthal kind of way call 'real' things.

A wide range of software programs, such as anti-virus tools and media players, can be downloaded from the CNET website. These have intrinsic technical features which can be measured and compared. Not quite the steel bars and lengths of cloth of the Industrial Revolution of the early nineteenth century, but their early twenty-first-century equivalents, which permit objective assessment of their relative quality. For anyone not inclined to carry out this task for themselves, there is plentiful and readily accessible information on the site in the form of both expert reviews of the different products and the opinions of ordinary users.

New Mexico-based researchers Rich Colbaugh and Kristin Glass did a study of real-life downloads of programs from the site to see if there was any way of predicting the daily totals for each of the different products, and presented their findings to a UCLA conference on complex systems in the agreeable setting of Lake Arrowhead in the San Bernardino mountains. They found that none of this information was of any use at all, whether it was a technical feature or the various reviews which were available. Colbaugh and Glass concluded that 'the average quality of the most popular software is not distinguishable from the average quality of all software available on site'.

A world in which the connection between the inherent characteristics of different choices and their popularity is broken is a completely different planet from the one the economically rational agent lives on, where incentives matter. In this rational world, we will try to sell more of our product by improving its quality and endeavouring to make its price more competitive.

We want to incentivise customers to buy more. But, it seems, in many modern contexts such a strategy may be pointless. A world in which network effects dominate incentive effects requires a radical reappraisal of how we behave.

In the software download study, at the start of each day we cannot predict using objective information which product in any particular software category will be the most popular download. All we know is that one of them will be the most popular (obviously), and that it will be downloaded many more times than the majority of similar products which are available.

The people who download a software product from the site during any given day will, in general, be completely unknown to each other. Yet they nevertheless make up a network. Users are given information on the number of downloads which have previously taken place during the course of that particular day. So a person choosing to download now may be influenced directly by the previous choices of others. He or she may download a particular program simply because it seems popular.

The precise composition of the network changes from day to day, in that the individuals downloading are not the same. But on each day, be it software downloads or YouTube hits or Flickr views, the structure of the network appears to encourage what we might term percolation across it. The choice of one or two programs, one or two videos, one or two collections of photographs, percolates across wide sections of the network. The most successful on any given day achieve huge popularity. The structure of the network seems to facilitate a small number of items – products, services, ideas, sites – receiving a great deal of attention.

*

A point discussed immediately below and to which we return at considerably greater length in Chapter 6, is that there are different kinds of networks, with radically different implications. Readers might usefully reflect at some point on the different networks in their own lives, of the different ways in which networks may influence their own decisions.

We can readily think of examples where networks are important in agents' decisions and where the networks, far from encouraging percolation, appear to be almost designed to resist the spread of different choices. Consider the world of economics textbooks.

Most students are fed not on esoteric maths but on the standard textbooks. But these have, if anything, gone backwards in recent years. Aimed at the mass market of US community college students, they have dumbed down the subject to a terrifying degree.

I have in front of me the 1967 edition of Richard Lipsey's *Introduction to Positive Economics*. This, along with Paul Samuelson's textbook, was the best-seller for many years. It is not aimed at geniuses, just ordinary, regular students, 'designed to be read as a first book in economics'. Of its 861 pages, only thirty-two contain any maths, and even that is of the simplest possible kind.

Yet it is full to bursting with really interesting examples of real-world behaviour. Yes, here is the basic model showing how in a simple market, price can adjust to bring supply and demand into balance. But here, too, is an immediate counter-example, of great practical importance, discussed at length. Indeed, it has its own separate chapter. What happens if supply can't be increased quickly, if it takes time to respond to price changes? (This is true of most agricultural markets: trees take time to grow, even chickens need five months before they can first start to lay eggs.)

Lipsey shows, simply and clearly, using only diagrams, how the free market might work very badly in this case. His chapter summary, printed in bold, states: 'in the unstable case, the operation of the competitive price system itself does not tend to remove any disequilibrium; it tends rather to accentuate it'. Careful, practical study is needed on a case-by-case basis to determine whether a free market is likely to lead to stable or unstable behaviour. The crude policy advice that markets always work is simply not given house room, even in a textbook for the ordinary first-year student.

So why are the textbooks not being rewritten, not just to bring back the insights of the word-rich, maths-poor texts of the 1960s, but to incorporate the advances which have been made in economics in recent decades, such as a more realistic, empirical view of how agents behave? Until I was drawn into the textbook world, this puzzled me.

A few years ago, I was approached by someone from a leading academic publisher. He was, he explained, their very top man across the whole of the sciences. His remit included economics. This sounded interesting. What did he have in mind?

What the commissioning editor had in mind was very exciting. He wanted an entirely new textbook, to incorporate not just the really interesting advances in the subject over the past twenty years or so but even to go beyond these into the world of networks. The editor already had a best-selling economics textbook of the standard kind in his stable. He understood that at some point in the future all existing textbooks will be redundant. The new generation of textbooks will contain the economics of the twenty-first century, not that of the twentieth (or even the nineteenth!), which the present ones do.

He was anxious that one of his rivals would get there before him, and bring out a textbook which would scoop the pool and be hard to dislodge from its number 1 slot. So he realised that his company would have to innovate and bring out a completely new text. Was I interested? It sounded like a dream. But like most dreams, it was too good to be true.

The editor faced a dilemma, which he articulated clearly. His problem was that the market – in this case the market for textbooks – is already occupied by the incumbents. They might ignore much of the interesting scientific work that has gone on in economics in the past twenty years, they might be guilty of dumbing down, but they are there. And their publishers and authors use every trick to make it stay that way. Top textbooks routinely have over 10,000 multiple choice questions helpfully provided on a web-linked site, for instance. So teachers don't even have to think about setting questions: they are all provided.

So we have a situation in which products with inferior qualities – containing lots of old-fashioned economics – are preventing products which are potentially superior from entering the market. The barriers to entry which they have erected are very hard to breach. In simple economics, this shouldn't happen. Consumers are supposed to have perfect information, so they should choose the new rather than the old. But the real world just doesn't work like this. Yet most students are now never told this, and they never get to choose – except by voting with their feet and dropping economics altogether.

And this was exactly the editor's dilemma. He knew that at some point the market will look completely different, that the new will eventually oust the old. But he had no way of knowing when this would be. In the meantime, any single attempt to enter the market with a new-style economics text would be likely to fail, unable to break the hold on the market which the current textbooks have.

We corresponded on this and talked. Eventually, the editor said he would go ahead on the basis that no more than 10 per cent of the total material could be the new economics, the other 90 per cent would be the old. But I just could not do it, I could not be part of disseminating a wrong-headed view of the world which leads to so much bad policy advice. I did not blame the man. He was thoughtful and anxious to do good, but faced commercial imperatives. Examples of textbooks which are trying to do twenty-first-century economics have since appeared, but their sales are very small regardless of their inherent quality.

In this example, it is the sheer density of the connections in the network which actively militates against change. Even the world of academic economics is not immune to innovation, far from it, and there is a chance, no matter how small, that their network would prove fragile. A completely new textbook could in principle see dramatic sales, as it cascaded across the network.

But each individual in a position to recommend adoption of the innovative text has so many strong connections which limit his or her powers to act. For lecturers there is the immediate pressure of being seen to act out of line with one's peers. There is the overarching network of the mainstream profession at the top level, which deems what is and what is not appropriate. There is the network which connects the lecturer to his or her students, who may or may not appreciate being taught differently from other economics students on their social networks. So it is very hard for any but the most determined to adopt the innovation. There are just too many connections which will not take the same decision, and the pressure to follow suit is very strong. Agents are more or less locked into a particular pattern of behaviour, which is hard to alter.

At the other extreme, we have the largely imaginary world of mainstream economic theory, in which individuals operate as entirely isolated agents. Choices cannot spread by contagion across the network, cannot cascade as a result of agents being influenced directly by the decisions of others, for the simple reason that there is no network at all.

In a universe such as this, we might start introducing a few connections, allowing at least some individuals to be affected directly by what others think and do. But, intuitively, when the agents remain only weakly connected in this way, the chances of a new idea, product, mode of behaviour, spreading across the network are limited. The potential for a cascade across most of the network is limited by the sparseness of the connections. Yet in the case of the economics textbooks given above, the potential is limited for quite the opposite reason, namely the sheer density and strength of the connections. These constrain any single agent from opting for a different choice from that made uniformly by its neighbours, using this latter word in the sense of the other agents to which it is connected. Everyone stays in line.

*

So, somewhat paradoxically, networked systems are resistant to change when they are either weakly or strongly connected. What happens when the degree of connectivity sits somewhere in between, when it is like Baby Bear's porridge, neither too hot nor too cold? It is in fact exactly the sort of mixture which is most likely to lead to positive linking.

Duncan Watts, whom we met briefly in the opening chapter, trained as a physicist at the University of New South Wales in Australia before moving to America, where he became Professor of Mathematical Sociology at Columbia University. More recently, Watts has moved to Yahoo! Research, where he directs the Human Social Dynamics Group. He explored the concept of cascades across networks in a brilliant paper published in the prestigious *Proceedings of the National Academy of Science* in 2002. The article has the austere and forbidding title of 'Global Cascades on Random Networks'. Although it is not exactly bedtime reading, it offers a simple but very powerful abstract model which tells us a great deal about the real world.

The description of his model, translating the maths into English, will take some time. But please be patient! The approach taken by Watts is similar to that used much more generally in network models. And the implications for policy are both surprising and profound.

Watts was interested in the question of what happens in a simple model in which, as a deliberate assumption, the *only* thing which affects how agents choose amongst alternatives is the choices which other agents have already made. In other words, in order to illustrate the potential impact of social influence, of choices being determined by what others do, he deliberately left out all other factors, such as price and quality.

Watts set up a computer model of individual agents who are connected to each other at random. We can usefully think of this as a game with some simple rules. One of the rules decides which agents are connected to each other. So we can choose to have, say, a hundred agents in the model* and decide that there is a 5 per cent chance of a given agent being connected to any other agent. On average, each agent will be connected to five others. This percentage can be varied each time the game is played.

In this context, the fact of being connected means that an agent to whom you are connected can potentially influence your behaviour. As we will see shortly, this does not mean that this agent will necessarily affect how you behave, but the small group to which you are connected are the only ones who have the potential to do so.

This way of connecting agents, by a purely random process, may seem entirely unrealistic but does in fact offer a reasonable approximation to many practical social and economic situations.

---

* In practice, there are usually a lot more agents in the model to avoid small sample problems, but a hundred is used to keep the arithmetic simple.

Epidemics are often spread by random contact. A person you do not know and will never see again sneezes on the train and you catch a cold. In a strange city, you choose the restaurant with more people in, even though you know none of them. In financial markets, a trader may very well monitor particularly closely the behaviour of a small number of others, but if the market starts to move strongly in one direction as a result of the decisions of many people entirely unknown to the trader, again a sensible decision might very well be to follow that trend, even if it is counter to what his 'control group' is doing.

Watts's game can be played with networks which have more explicit, much less random social structure to them, of which more later, but let us describe the rest of the rules of the game retaining the assumption of a random network. In this model, an agent has a choice between two alternatives. These could be a consumer deciding between two competing brands, a firm considering two different technologies, someone considering in the England of the 1550s whether to remain a Protestant or become a Catholic, to give just a few examples.

In reality, people will take into account a whole range of factors in making these decisions, but in all these cases, no matter how much information is used to make the choice, by assumption we are dealing with either/or. There may of course be more than two choices (including not choosing either of the alternatives on offer), but this can readily be accommodated in the model, and we concentrate on the simplest version of the model where the choice is between two alternatives and by assumption the agent chooses one or the other of them.

When the game starts, by assumption all agents have chosen alternative A. We need now to specify a rule of behaviour which determines whether they stay with A or switch to B.

We first of all make the entirely realistic assumption that each agent differs in his or her intrinsic willingness to switch. Take consumer products: some individuals are keen to try new products, whilst others have a preference for staying with what they already know and like. The more information we have about the persuadability of agents or their willingness to experiment, the more realistic the model can be made. But for the moment imagine we have no information on this at all. Lacking any better alternative, we can simply allocate at random to each agent a value between 0 and 1. Slightly confusingly, an agent allocated a number close to 1 is deemed to be *less* persuadable, *less* willing to switch than someone allocated a number close to 0. The reason for this will become clear. For purposes of description, we call this value the agent's *threshold*.

How, then, do agents decide whether to switch from A to B? In this game, by assumption the only information used by the agent in making this choice is the choices which the other agents to which he or she is connected have also made. Incentives do not enter the picture. If the agent is in state A and the proportion of these relevant agents who have chosen B is above the agent's threshold, the agent will also choose B instead of A. So if your threshold is 0.5, say, and three out of the five agents to which you are connected have chosen B, you will switch, because 3/5 = 0.6, which is greater than 0.5. But if only two have chosen B, you stay with A. 2/5 = 0.4, which is less than 0.5. It is apparent now why a higher threshold means that the agent is less persuadable than an agent with a lower threshold. Someone with a threshold above 0.8 will need all of his or her network to choose B before being persuaded to switch, whilst if it is less than 0.2, even just one person choosing B will lead the agent to also make this choice.

As noted above, there may be many factors which an agent takes into account in deciding between A and B. The choices made by those people whose opinion or behaviour he or she respects may very well be one of them, but not necessarily the only one. A simple example is if A and B are competing consumer brands, their prices may also be an element in the decision as well as what other people have chosen. But the essential features of Watts's model continue to be valid as long as the choices of others remain a key factor. Other things such as price can again be accommodated in the model; indeed, it is always easy, if not always edifying, to make models more and more complicated by bringing in more and more factors.

Besides, the assumption that the behaviour of others is the only factor may often be a reasonable approximation to reality. In the restaurant example above, you may have a guidebook to the city which has enabled you to filter down the options to just two, but the number of people in each may still be the decisive factor in your choice. In situations such as this, you have relatively small amounts of information on which to make a judgement, so relying on the choices made by others makes sense. In other situations, people may be able to acquire large amounts of information about products which are inherently difficult to understand. Processing and understanding the information available on, say, a choice of pension plans is a hard task. So a reasonable decision rule is to rely on the actions or recommendations of a small number of people whose judgement you trust.

*

Thomas Schelling is an American polymath who won the Nobel Prize in economics in 2005. His work ranges across not just economics but areas such as game theory, foreign affairs, conflict resolution and nuclear strategy. Back in 1973 he published an article in the rather obscure *Journal of Conflict Resolution*. The title of the paper is much more memorable, if somewhat bizarre: 'Hockey Helmets, Concealed Weapons and Daylight Saving'.

But that is not all, for it goes on to add, in smaller type, 'A Study of Binary Choices with Externalities'.

This latter phrase defines what has become a huge area of study in network theory, including Duncan Watts's illuminating model. What does it mean? 'Binary' means involving two things, so choices are situations such as those described above, where an agent faces a choice between two alternatives. 'Externalities', a very useful concept in economics, are situations where the decisions of any one agent can have consequences for others. For example, a factory which emits pollution creates a negative externality for inhabitants of the area. The factory produces goods which it sells, employs people in the process of doing so, but at the same time imposes costs on everyone else as a result of the pollution it creates. Unless there is some form of taxation on pollution in place, these costs are external to the firm itself, and have to be borne by other agents.

The concept has been an important one in economic theory for at least a hundred years, and there is a huge literature on the topic. Schelling incorporated the idea into networks for the first time. His inspiration was an event which took place in American ice hockey in 1969. A leading player, Ted Green of the Bruins, was not wearing a helmet and in a clash suffered a fractured skull. Schelling's paper begins with a quote from *Newsweek* on the incident: 'Players will not adopt helmets by individual choice for several reasons. Chicago star Bobby Hull sites the simplest factor "Vanity". But many players honestly believe that helmets will cut their efficiency and put them at a disadvantage, and others fear the ridicule of their opponents . . . One player summed up the feelings of many: "It's foolish not to wear a helmet. But I don't – because the other guys don't."'

I don't because the other guys don't! A short phrase which captures much of what modern network theory is about in social and economic situations. Here, the player actually *has* made an independent, 'rational' assessment of the costs and benefits of wearing a helmet and has concluded that it makes sense to do so. But the network effect, the impact of peers on the behaviour of individuals, trumps this calculation.

This unfortunate incident inspired Schelling to write the paper. As he points out 'the literature on externalities has mostly to do with how much of a good or a bad [e.g. pollution] should be produced, consumed or allowed. Here, I consider only the interconnectedness of choices to do or not to do, to join or not to join, to stay or to leave, to vote yes or no, to conform or not to conform to some agreement, rule or restriction'. Nearly three decades after the publication of the Thomas Schelling paper, Duncan Watts was inspired to formalise the concept of binary choice with externalities and to explore its implications more deeply.

*

So with all this in place, we are now in a position to play the game, or, more scientifically speaking, to run the model. Initially, remember, by assumption everyone has selected option A. The game is started by choosing a small number of agents at random to switch to B. Imagine that we have some sort of policy which induces this behaviour, some sort of nudge factor, some incentive, which succeeds initially in altering the behaviour of only a few people.

The purpose of the game is to see how many agents eventually end up selecting B rather than A. The process by which they do this is defined by the 'copying rules': who you are connected to (i.e. who can potentially influence your behaviour), how persuadable you are, and how many of your potential 'influencers' are making a choice different from your own. In turn, if you are persuaded to switch from A to B, you will potentially influence people who look to you as part of their decision-making processes to also switch.

The result of any particular 'play' of the game may be very sensitive to the particular circumstances. At one extreme, suppose the agents who were selected to make the initial switch from A to B were connected to agents who were very hard to persuade, who required almost everyone who might influence them to choose B before they themselves did. The 'cascade' – the spread across the network of people choosing B rather than A – may very well be stopped there and then. No one else chooses B at all beyond the small group assumed to do so as a result of the initial change in incentives which we might readily imagine in a real-life situation encouraged them to make this initial switch.

In practice, the more information we have about the agents, who they are connected to, how persuadable they are and so on, the more we can start calibrating the model to a real-life situation. But in the very general abstract way in which Watts played his game, such information is necessarily lacking. This, it must be stressed, is not a defect of his method but a strength. By exploring a wide range of initial choices, by having them connected to different sets of other agents, by giving these different levels of persuadability, we can start to understand the general properties of the model across a wide range of assumptions.

To do this in practice, the game is played many times under identical rules. The only difference in each solution of the model is the agents chosen at random to switch to B at the very start.* A crucial point is that the size of the initial disturbance, the initial shock to the system, is exactly the same in each solution. The same number of agents is selected to switch from A to B each time.

* Theoretically, of course, given that they are chosen at random, these could be identical in two separate solutions, but the chances of this are vanishingly small.

Examples have already been given of the sort of real-world settings which this model might help illuminate. There are others. The network might be the power grid of the United States, how power is transmitted across the country. State A means that each generator is working well, state B means that it has failed. A small number of outlets chosen at random experience a failure, the sort of thing which happens all the time. How far will this spread? In a different context, sentiment about the future, the degree of optimism or pessimism which firms feel at any point in time, is an important determinant of the boom and bust of the business cycle. Here, we are in Keynes's world from an earlier chapter, where we have 'a society of individuals each of whom is endeavouring to copy the others'. We can think of a firm in state A as being optimistic. The economy receives a small shock, a bit of bad news, and a few firms switch to state B, pessimistic. How many others will abandon their optimism? If enough do so, the economy will move from boom to bust. But by assumption, the economy in this case has received only a small adverse shock. Can this really be sufficient to precipitate a full-blown recession?

The answer is both yes and no! The same small initial disturbance can have dramatically different outcomes. Most of the time, the initial disturbance, the initial switch by a small number of agents from A to B, does not spread very far. But occasionally there will be a cascade across the system as a whole, and most agents will end up with B instead of A.

In the above few paragraphs we have fleshed out a feature of some networks highlighted in the first chapter – systems of interconnected agents whose behaviour influences each other are both *robust* and *fragile*. Most of the time, the system is robust to small disturbances, they do not spread very far. But occasionally, the system is fragile, vulnerable to exactly the same size of shock which it is usually able to contain. These properties present both difficulties and opportunities to policy makers.

Figure 5.1 shows the results of 1,000 separate trials of the model,* and the distribution of the proportion of all the 1,000 agents in each who eventually switch from A to B.

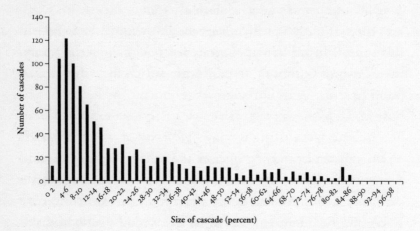

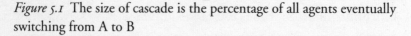

Size of cascade (percent)

Figure 5.1 The size of cascade is the percentage of all agents eventually switching from A to B

Note: The data is grouped into bands of 2 percentage points, so the first bar on the bottom axis shows the range 0% to 2%, the next 2% to 4% and so on

Out of the total of 1,000 solutions, the vertical axis indicates how many of them were in a particular range and the horizontal axis shows the range. The largest bar shows that on some 120 occasions out of the total of 1,000, the percentage switching to B was small, in fact in the range of 4–6 per cent the way we have plotted the data. Next, we see around 100 solutions ending up in each of the ranges 2–4 and 6–8 per cent, and around 80 in the range 8–10 per cent. So most cascades are small, the initial

* Technically, this example is of a random network with the probability of connection set at 4 per cent, and ten initial seeds.

disturbance to the system when a few agents switch to B does not spread very far. The system is robust to shocks. But we also see a few occasions when there are very large cascades, over 80 per cent of all agents, in fact. It is therefore at the same time fragile. And, importantly, it is an entirely random process that 'decides' whether the *same network* is robust or fragile.

There are many subtleties even to this simplest version of the Watts model. But its implications for policy, in circumstances where network effects matter, are both disturbing and exciting.

*

If the world operates in anything like the same way as it does in the model, anticipating the impact of a change in policy becomes extremely difficult. The common-sense causal link between the size of an event and its eventual impact is broken. Of course, if a large shock were administered to the system so that, say, one half of all agents switched from A to B, by definition the eventual outcome would be large. But, equally, a small disturbance can have dramatic consequences.

However, by deliberate construction in Watts's model, *all* the shocks administered to the system – the number of agents select-ed to switch from A to B at the start of each solution – are the same. Yet the outcomes, the proportion who eventually switch to B, can be dramatically different.

This highly counterintuitive result is disturbing. How can it be that a small change can have a massive consequence? An ini-tial reaction might well be that this is in some way an artefact of the model, which makes many abstractions, many simplifications from the real world. Surely human societies and economies sim-ply do not operate like this?

But, as we have seen, they do. A man ironing clothes on a closed motorway gets 4,000 times more people watching his video than a man ironing clothes on the summit of Britain's highest peak. It is hard to see how there is anything intrinsically more interesting, more attractive in the former activity rather than the latter. With a dramatic switch in scale and scope, we can recall the 20 per cent collapse of stock market prices on that Black Monday in October 1987. Faced with a truly major event such as this, instinctively we look for the smoking gun, for the massive event which triggered it. An intensive hunt has been mounted, but the culprit has never been found. Traders on stock markets receive large numbers of potential shocks in the form of new information, whether about the overall economy, particular firms, or the actions of other traders. Each piece of new information has the potential to trigger a large cascade. Few do. For the most part, the disturbances are contained by the robustness of the network. Every so often, the system proves fragile.

In some ways, this is good news for policy makers. Suppose a desirable policy aim is selected, such as reducing the number of people who are obese. Policy instruments are chosen, which might include good old-fashioned incentives through tax increases on fatty foods, as well as less direct methods such as health education, restricting advertising, or whatever. Now, to achieve a big reduction in obesity, the use of incentives alone, no matter how smart or sophisticated, requires that the policy has a big effect, that it alters the behaviour of large numbers of people. Incentives plus networks means that, if you are lucky, the behaviour of only a small number of people needs to be changed, yet the number who eventually change their minds could be enormous.

This represents a potentially huge increase in the ability of policy to affect outcomes, to reap the benefits of positive linking. But in a networked world, things are rarely as clear cut as that. Suppose some individuals were indeed induced by a change in incentives, by a straightforward change in price or by some more subtle factor to alter their behaviour in the way intended.

However, the perception that the authorities were trying to influence people might induce others, through the network effect, to become more stubborn or even to adopt a completely contrary mode of behaviour. We have seen examples of this already, not least the experience of the Italian police captain confronting the English thugs and firing his pistol into the air.

Moreover, and more generally, networked systems bring problems to policy makers trying to evaluate the effects of previous policies. What worked and what did not work? A great deal of policy evaluation is carried out paying little or no attention to the potential impact of network effects.

But if these are important in any particular context, studies which ignore them can generate quite misleading results. A successful outcome may arise, not principally because of a partial, initial success with a change to incentives which leads a few agents to alter their behaviour, but because of the impact of imitation across the network. In such a case, the success would be mistakenly attributed to the incentive factor, and policy makers would be puzzled when a similar policy led to an apparent failure in a different context. In the marketing world, for example, successful viral marketing campaigns, whose specific purpose is to spread across a network, are notoriously difficult to repeat. The creators of rather spectacular successes find them hard to replicate.

The difficulties in identifying whether incentives have worked in the past, and to what extent, can be seen in an example that might be found in any basic introductory course in economic theory. We have a group of individuals contemplating whether or not to buy a particular product. Each individual has his or her own intrinsic preference for what is on offer, so each will decide to buy at a different price. Some are strongly attracted and will pay a high price, others will buy only if the item is perceived as cheap.

The usual interpretation of price is, of course, exactly that.

So we might examine a brand of shampoo and see how its price affects sales. But as we have seen, 'price' can have a much wider, multidimensional interpretation. It essentially summarises the costs associated, or thought to be associated, with any particular course of action. So, yes, it can just be the price of your favourite shampoo. But it could be the perceived costs associated with, say, being a petty criminal, taking drugs, or being burned to death.

*

From the individual preferences, the prices at which different people will buy the product (carry out the activity), we can easily obtain a 'market' demand curve. In other words, we add up the individual decisions and see how much is bought, how much of the activity is carried out, at different prices.*

With Amy Heineike, then of George Mason University, I investigated what happens to the very simple demand curves of economic theory when network effects are present in the system. In this very basic model, the top left-hand chart in Figure 5.2 represents the classic market demand curve. As price increases, demand falls, exactly as expected. If we can discover the shape of this curve by, say, some smart statistical analysis of the data, we can change incentives – the price – to change the amount people buy.

We then introduce into this elementary model what in this context we term the 'bandwagon' effect, so that the more people buy the product at any given price, the more likely any given individual is to buy it as well. The additional charts show the overall demand for the product (degree of participation in the activity) with different strengths of the bandwagon effect. The stronger this effect, the less price matters.

* For economic theorists, I am ignoring here any Sonnenschein–Mantel–Debreu effects which cause such fundamental problems in principle for the scientific nature of general equilibrium theory.

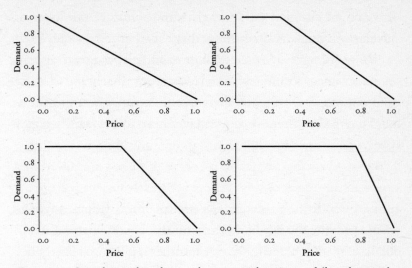

Figure 5.2 Simple market demand curve with price and 'bandwagon' effects. In the top left-hand chart there is no bandwagon effect, and demand simply depends upon price. The other three charts show the impact on demand of introducing stronger and stronger bandwagon effects.

Again, to repeat for explanation, the top left-hand chart shows the standard demand curve of the economics textbooks. As price increases, less of the product is bought. On the left-hand axis, labelled 'demand', we are plotting the proportion of the total number of people who are both interested in the product and actually buy it. So at a very low price, the proportion is close to 1, in other words almost everyone who might want to buy it, does buy it. As the price increases, the proportion falls, until eventually, when the price is sufficiently high, no one buys it at all.*

There are many questions from a policy perspective even with this simple chart. How do we know what the relevant measure

* The scale for price, between 0 and 1, is quite arbitrary for this illustrative example. Readers can imagine for themselves the actual price range over which either everyone or no one who might be interested would actually buy a product.

of 'price' is? How do we know the distribution of the inherent preferences of agents about the activity and hence how they react to changes in price? How do we know which other agents' actions are taken into account by any given agent?* But many of these questions apply even when there is no network effect present at all. They reflect the difficulties and uncertainties which policy makers face even in an apparently simple world.

But suppose that somehow all these problems are solved in a reasonably satisfactory way. We can see the challenges and opportunities which the existence of network effects brings. Imagine we are near the top left-hand corner of the chart. Participation in the activity is high, the costs associated with it being small. Policy makers want to discourage this form of activity and so increase the price, again using 'price' in the general sense of the term. It could, for example, refer to a more punitive criminal justice system in which the 'price' to the individual of breaking the law is the increased probability of being sent to prison for a longer period.

In a non-networked world, in the top left-hand chart, it is easy to see whether or not the policy is working. Put the price up, and demand falls. But if agents base their actions in part on the actions of others, increasing the price initially has no effect. Then suddenly, as can be seen in the other three charts, we get not just a reduction, but for any further small increase in price, we get a bigger change in demand than would take place in the absence of network effects. By this stage, however, the authorities might easily have concluded that the policy of increasing the price had not worked well before this critical point was reached.

Studies of past changes in prices which attempted to estimate

* This chart assumes agents know and react to the actions of all other agents, so the analysis retains an important feature of conventional economic theory with respect to the information set available to agents.

the impact of price on demand without taking into account net-work effects might also provide very different stories to policy makers, depending upon the part of the chart from which the evidence was taken. Often, with evidence taken from only a lim-ited range on the chart, the policy would show no effect at all.

But the reverse starting point, near the bottom-right corner of the bottom-right chart, shows the huge potential gains to policy makers in trying to encourage certain types of behaviour – buy-ing the product – if network effects are strong. Even a small reduction in price might have a powerful effect.

*

The crucial challenge for policy makers is to understand and take account of the fact that networks are becoming more and more important in the social and economic world. The internet revolution in communications technology is obviously a key fac-tor. But the entire second half of the twentieth century featured the massive rise of globalisation, a huge increase in travel, and a greater and greater proportion of the world's population living in cities, exposed to many more people, many more networks than they would be in the confines of the village. The model described above is almost as simple as you can get, but it creates both opportunities and problems for policy, for positive linking.

The first problem for policy makers is simple: how do we know whether network effects are important in any given context? Do we have to spend a huge amount of time and resources investi-gating this issue, do we need more-or-less complete information, before we can even start to think about the policy implications? Or is there a Herb Simon-like rule of thumb which enables us to detect the presence of network effects without too much effort?

The second problem relates directly to the elementary market demand curves described above. Even if somehow we know that networks matter when we are considering a particular issue, they seem to introduce layers of additional uncertainty into trying to work out in advance what the effect of any given policy change might be. Gathering accurate information on the demand curve might be difficult in practice, even without network effects being present, and assessing the impact of changing incentives may often be hard, as we have seen in many examples. But in principle these problems can be overcome.

Networks appear to make things even more difficult, even more challenging for the policy maker. So much seems to be due to chance and contingency. To what we call in plain English, 'luck'. Are there ways of trying to reduce this uncertainty, of formulating new guidelines to offer policy makers to enable them to benefit from the new insights, the positive linking, which networks provide? These are the questions we consider in the next chapter.

6

The World Is Not Normal

Standard economics has demonstrated the importance of incentives. Behavioural economics establishes that these operate rather more subtly than conventional theory suggests, but still remains an analysis of incentives, from which network effects are notable by their absence.

We have already seen practical examples of the effects of networks, which alter the impact of incentives, whether standard ones such as price or the wider set identified by behavioural economics. And we have seen that where they are important, they introduce even more potential uncertainty into the outcome of any particular policy change than already exists when we think just about incentives and the ways they might work. Networks, in which agents copy or imitate the behaviour of others, can either enhance the effects of incentives, in whatever form, or completely swamp them. Positive linking is a very powerful force. This uncertainty is an inherent feature of the world, which no amount of cleverness will eliminate. But there are ways of getting to grips with this uncertainty, of obtaining practical guidance on what to do, on what might work, in situations where network effects matter.

But first of all, we need to ask: is there some way of knowing whether network effects are important in any particular situation? Fortunately, there is. We are not left completely floundering in the dark. There is a readily identifiable piece of evidence to be found wherever network effects are important. Decision makers can be alerted to the fact that the opportunities exist for positive linking.

A great deal of quantitative work in economics is based on a concept developed over 200 years ago by Carl Friedrich Gauss, one of the greatest mathematicians of all time. Statisticians are very interested in how the members of any particular group of objects, for example the population of cities in the United States, are distributed. In this context, 'distribution' does not mean the physical location of the cities. Rather it refers to how the sizes of the various cities are spread across the group as a whole. How many have a population greater than a million, how many between half a million and a million, or whatever bands of size we choose to select.

Gauss worked out the properties of a particular kind of distribution. It is often described, not surprisingly, as Gaussian. But, just like New York City, it is so good that it has been named twice. Its other name is 'normal'. An everyday illustration is the heights of individuals. The average height of American adult men is 5 feet 10 inches. It is also true that more men are of this particular height than any other. There are similar, though slightly smaller, numbers close to the average on either side, who measure 5 feet 9 or 5 feet 11 inches. The number of men we observe who are of any particular height will be less and less the further we move from the average. The resulting distribution looks like the one in the upper panel of Figure 6.1 on p. 160. This is just one example of many such distributions that exist in reality, and because it is common in nature, we call it 'normal'.*

The normal pattern we observe in the heights of American

---

* There are subtle arguments about whether this height distribution is exactly Gaussian, but everyone agrees that it either is, or is very close to being so.

men has two features. First, it is symmetric around the average. The number of men who are 5 feet 9 inches is very similar to the number who are 5 foot 11 inches, and so on. Second, and crucially in this context, in a normal distribution very large differences from the average are never observed in practice. No one has ever been found to be 10 feet tall, or 3 inches.

The measure statisticians use to describe how the numbers fall away as we move further and further from the mean is something called standard deviation (signified by the Greek letter sigma (σ), as readers familiar with statistical theory will know). Two-thirds of all the observations of data that follows the normal distribution will lie within one sigma of the average. In the case of American males, a so-called 'one sigma' event is just three inches. So two-thirds of American men are between 5 foot 7 inches and 6 foot 1 inch tall. That is a lot of people: roughly 80 million, to be slightly more precise. But in a normal distribution, these numbers drop very rapidly as we move further away from the average. Michael Jordan, probably the greatest basketball player of all time, at 6 foot 6 inches, is a 'three sigma event', and only 0.15 per cent of American men are taller. A very small percentage, but still a sizeable absolute number, some 150,000. Yao Ming, a contemporary star, at 7 foot 6 inches is a 'six sigma' event, and people of this height are so rare that he had to be imported from China.

*

The normal distribution underpins almost all the technical statistical analysis carried out by economists. Herb Simon, the great polymath we met at length in Chapter 3, looked at the world, and noticed that across a very wide range of examples, data was distributed in a completely different way. Simon was not the first person to notice that each of these examples were very definitely not normal distributions. But he was the first to pull together quite disparate examples of the same phenomenon.

He noted that 'its appearance is so frequent, and the phenomena in which it appears are so diverse, that one is led to the conjecture that if these phenomena have any property in common it can only be a similarity in the underlying probability mechanisms'. His examples were: distributions of words in English prose by the frequency of their appearance; distributions of scientists by numbers of papers published; distributions of cities by populations; distributions of incomes by size; distributions of biological genera by numbers of species. A diverse set indeed.

A key difference between Simon's data and data distributed in a Gaussian, or normal, way is the huge degree of inequality of outcomes which we observe. Unlike in the normal distribution, there are massive deviations from the average.

Such deviations are described by the word 'skew'. This was a key word in the enticing (!) title of Simon's path-breaking article on the topic, published in *Biometrika* in 1955: 'On a Class of Skew Distribution Functions'. The basic concept may seem austere, but it features in the opening sequences to one of the Monty Python sketches on the Spanish Inquisition. Set in the early years of the Industrial Revolution, the elegant daughter of the factory owner is sat quietly in the luxurious family home. Her obviously upper-crust fiancé bursts in with disturbing news from the factory. 'One o't crossbeams gone askew on't treadle', he pronounces in a broad northern accent, faithfully reproducing the mill-worker's report of the incident. Despite repeating the phrase, this time in his usual upper-class accent, neither of the pair has the slightest idea what it means. We can assist. A piece of machinery has 'gone askew', meaning it has become heavily displaced. In skewed distributions, there are observations which are heavily displaced – a long way away – from the average value. The greater the skew, the more extreme the biggest values will be.

Here is a more everyday example of skew. The first three sites to come up on a Google search attract 98 per cent – ninety-eight per cent! – of all subsequent hits from the search. *All* other sites popping up in the search get just two per cent of the traffic. This is a truly massively skewed distribution. Even within the top three, there is striking inequality, with the number one site receiving 60 per cent of all hits. The typical household income in Britain is some £25,000 a year. In America it is around $50,000. But there appears to be almost no limit to the amount by which some people's incomes exceed the average. Hedge fund owners will often pay themselves annual incomes in the tens of millions, whether pounds or dollars, and there are several examples of individuals who have been paid – I do not use the word 'earned' here – more than a billion dollars in a single year. A billion dollars!

Before going on to give examples of non-normal outcomes in a very wide range of human social and economic activities, a chart might be useful to bring out the differences. Figure 6.1 illustrates a normal and a typical non-normal distribution of data. The charts do not represent any particular examples, but the scale of the data in each chart is directly comparable. In both the charts, the *average* value of the data is the same. But the ways in which the values are spread around the average are completely different.

In both cases, there are 10,000 data points, with an average value of 7.5. We divide the data up into small ranges, and see how many of the data points lie within each range. So in the first chart, the tallest bar in the chart – just – sits exactly round the average value of the data. This shows how many of the data points in a normal distribution with an average value of 7.5 lie in the range 7.45–7.55. Reading across to the left hand side, we can see that there are just over 400 of them. Close to the average, there are also lots of other data points. But as we move further away from the average, on either side, we can readily see that the

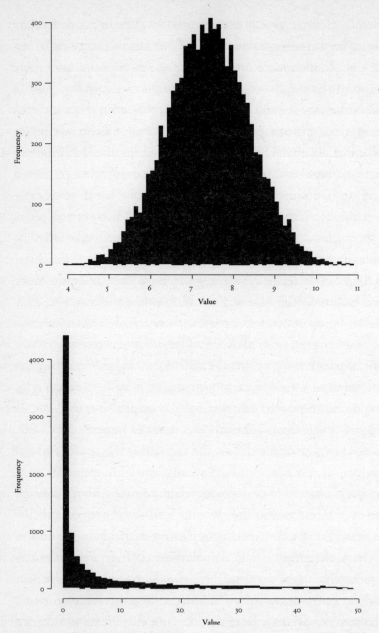

Figure 6.1 Typical examples of normal and non-normal distributions. By construction, the average value of the data is the same in both panels.

THE WORLD IS NOT NORMAL

number of times we observe data with such values falls away quite rapidly. There are very few data points with a value of less than 4.5, and a similarly small number which take a value of more than 10.5.

The non-normal distribution is very different. It is not even apparent that the average value of the data is 7.5, although readers can be assured that it is. What we see here is a huge number of data points, well over 4,000, which take a value of close to 0. And we see a small number which have values of more than 50. So it is not just the shape of the distribution which is different, but the spread, or skew, of the data. In the normal distribution, most observations are confined to a narrow range around the average, and there are no really large deviations from this. The ratio of the highest value to the lowest is around three. In the non-normal, there is a massive spread, with the ratio of the highest (a value over 50) and the lowest (close to 0) being around 1,000!

Most urban agglomerations, be they towns or cities, are relatively small in terms of their populations. Tertius Chandler has compiled a marvellous book, *Four Thousand Years of Urban Growth: An Historical Census*. The contents are exactly what it says on the label. Estimates are given of the population of a large number of cities across the world over a period of four millennia. So we learn, for example, that in 612 BC Babylon was not only the largest city in the world, but was the first to ever have a population of over 200,000. And this was at a time when most human settlements were tiny, with populations probably numbered in single figures. Babylon dwarfed them, but there was only one Babylon. There are of course all sorts of arguments as to where exactly a city's boundaries lie. On the old city limits, the current population of Tokyo is only (!) around 9 million, but the urban mass which constitutes the Greater Tokyo area has a population of 35 million. Either way, the figures overwhelm the large number of towns around the world with populations of, say, 10–20,000. There are a few very large cities, and lots and lots of very much smaller ones.

In the last decade or so, many more examples of this kind of skewed distribution have been discovered. For example:

* downloads on YouTube
* film producers' earnings
* the number of sexual partners people have
* the size of price changes in financial assets
* crowds at soccer matches
* firm sizes
* the size and length of economic recessions
* the frequency of different types of endgames in chess
* the ratings of American football coaches in *USA Today*
* the distribution of £1 million homes across London boroughs
* unemployment rates by county in America
* deaths in wars
* the number of churches per county in William the Conqueror's Domesday Book survey of England in the late eleventh century.

There are arcane disputes about which variety of 'skew distribution' best describes each of these examples – remember the words in the title of Simon's article: a class of skew distributions, meaning not symmetric at all, heavily skewed in one direction or the other – but they all exhibit very marked degrees of inequality of outcome.

*

The idea that highly unequal outcomes are pervasive in the human social and economic worlds is interesting in its own right. But such outcomes are also of great practical importance for the policy maker. When we observe them, we know that network effects are almost certain to be present in the behaviour of the agents who constitute the system in which we are interested. As we will see below, the basic mathematics of how network effects operate enable us to make this statement. The agents are not merely reacting to incentives. They are changing their behaviour directly in response to others. They are copying, imitating the opinions, choices, behaviour of other agents. Incentives may also still matter, but network effects dominate their impact.

Non-Gaussian outcomes are the classic signature of network effects. This is obviously a very useful piece of knowledge for policy makers. Wherever we see such outcomes, the possibility of taking advantage of positive linking also exists. We do not need to invest huge amounts of time and resources in discovering whether network effects are present. We simply look at – more likely, get one of our advisers to look at! – the distribution of outcomes. This is a powerful rule of thumb with which to detect the presence of network effects. Unfortunately, most policy makers with training in the social sciences, notably in economics, have been taught to assume the Gaussian distribution.

Of course, it is one thing to make this point, another to demonstrate why this should be the case. To do this, we will have to delve in some detail into Simon's skew distribution paper. But before girding our loins for these rigours, we might usefully reflect on many of the examples we have encountered and ask how these might arise in a world where only incentives mattered, a world in which network effects were absent.

Remember the eccentric Englishmen featured on YouTube videos, one of them ironing on the top of a Scottish mountain, the other ironing on a temporarily closed section of motorway. One proved 4,000 times more popular than the other. It is very hard to imagine that the inherent preferences of those who viewed these videos were skewed in this dramatic way. Indeed, we can get a rough estimate of the relative inherent interest in the two by doing a search on the Google UK site. The word 'motorway' is used in British English to describe roads built specifically for higher-speed traffic. In other languages, or other variants of English, it is known as 'freeway, autoroute, Autobahn, autostrada' and so on. The point here is that 'motorway' is a specifically British way of describing this type of road, so a search on the UK Google site is the relevant one. If we search 'motorway', we get 30.2 million sites, and 'Ben Nevis', the mountain where the ironing was done, yields 2.9 million sites. So, as an approximation, 'motorway' has ten times the level of inherent interest as 'Ben Nevis'. But the former video attracted 4,000 times as many hits as the latter.

Another internet-related statistic was cited above. The first three sites to come up in a Google search typically attract 98 per cent of the subsequent hits by the searcher on that topic. The principle of behaviour which underlies these examples is not in any way confined to the internet, it is present in every single one of the examples quoted and discussed in the book. But the internet examples are both everyday ones, and enable us to focus on what this process, this principle, is.

Once a site, a video, a photograph starts to become popular on the internet, it becomes more popular simply because it has already become popular. A site gets enough hits for its head to start to poke out above the parapet, to impinge on the consciousness of users of the internet. And it is then very likely to get even more hits, purely because people have become aware of it, rather than the huge number of other sites which exist, even when a particular topic is being searched for. Almost every single reader of this book will be familiar with the behaviour being described here. You might open a page of a newspaper early in the day, or visit one of your favourite sites before getting down to work. An item catches your eye in the 'trending now' category. You had absolutely no intention of looking at anything to do with this particular issue when you opened the site, you may even not have been aware of its existence. Yet you may click on it – not the same at all as saying you *will* definitely click on it – simply because it is 'trending', because at that moment it is popular.

It may not be immediately obvious that when you are distracted in this way, you are being influenced by a network. But it is indeed a network which is the main driver of your behaviour when you make that click on to the 'trending' item.

It is the network of all the people who have previously clicked on the item. For the most part, any single individual who has done so will be and will remain completely unknown to everyone who subsequently clicks on the same thing. But he or she is part of the network. The network which influences you, in your turn, to investigate the same item. Your behaviour has been directly affected by the decisions of others, even though in this example they are anonymous to you. They, and now you and others who in the next few hours may be influenced by your own decision to make the click, are temporarily connected on a network, the network of those who clicked on the same site.

The network here is temporary. It will evaporate as other items take its place. But in other instances which we have seen, networks may last much longer, such as the network across which the Protestant martyrs of England influenced people's behaviour. Again, the internet examples are used for illustration here because they are such familiar occurrences. It is not every day that you reflect on changing your political ideology or religious belief. But lots of people use the internet on a regular basis for much less momentous purposes.

We can now return to Herb Simon and his 1955 paper in

Biometrika. Simon did not discover the empirical examples about which he wrote, which showed that non-Gaussian outcomes of human social and economic processes are so widespread that these, and not the so-called 'normal' distribution, appear to be typical of our world. The originality of his paper was in the fact that he articulated effectively an underlying mathematical process by which such outcomes can be generated.

The process was rediscovered in the 1990s by physicists such as Albert-László Barabási of the University of Notre Dame, who gave it the name of 'preferential attachment'. As we have seen above, it is based on the principle of 'to him that hath, more shall be given'. The twist is to add a few words: 'to him that hath, it is likely that more shall be given'. This is the fundamental building block of behaviour which underpins positive linking.

All scientific theories involve making dramatic simplifications and, especially in the social sciences, they are approximations to reality, not the same as reality itself. This point cannot be stated too frequently. So Simon's process is not intended to be a complete description, the whole story, of why we observe some particular outcome. But it is a mechanism which gives us an insight into many qualitatively similar outcomes from widely different fields.

Simon did not say that someone making a choice would necessarily choose the most popular item, opinion or whatever. But that it is more likely that an agent will choose this than a less popular alternative. The basic idea is straightforward. Suppose there are just three choices available to you, whatever these may be, and you are wondering which one to select yourself. One has been already chosen 6,000 times, one 3,000 and the final one just 1,000 times, making a total of 10,000 altogether. If we assume for purposes of illustration that the only rule of behaviour you are using when making your choice is that of preferential attachment, the rule says the following. You may actually choose any one of the three alternatives. But you are twice as likely to select the most popular rather than the second most popular, and six times as likely to choose this as the least popular. The most popular has been already selected 6,000 times compared to the second most popular, twice as many, so you are twice as likely to select it. And it has been selected six times more often than the least popular.

At first sight, this may seem a strange way of behaving. You are paying no attention to the attributes, to the features of the three alternatives. But the top three sites which are followed up on a Google search typically reflect exactly this pattern. The three of them get almost 100 per cent of the subsequent hits after the search, and the top one of them all gets 60 per cent of the total. Again, scientific theories are approximations to reality, but as a description of behaviour in a complex environment, preferential attachment seems to have something going for it.

*

This long discussion has been trying to put substance on the assertion made above that there is a typical 'signature' of social and economic outcomes in which network effects have an important influence on how agents behave. We simply inspect whether the various choices – whether opinions, consumer goods, types of behaviour or whatever – are selected in numbers which are 'skewed' or whether they are more like a Gaussian distribution. If we see a few choices being selected in relatively large numbers and most choices being made by few, we know in all likelihood we are in a networked world.

Network effects, as we have seen, introduce even more uncertainty for the policy maker in terms of assessing in advance – or even after the event – what the impact of any particular policy might be. But at least we have a straightforward way of knowing whether network effects are important in any given context. At least we can alert the policy makers to both the potential difficulties these create, and the benefits of positive linking which can be potentially exploited.

We can do even more. So far, we have simply spoken of 'networks' in a generic sense. But there are different types of networks, in the same way that there are, say, different types of mammals. We can think about 'mammals' in general, or we can think specifically of, say, rabbits or lions. And the more specific we can make the description, the more useful it is. Each individual rabbit or lion will differ in various ways from others of the same species. But the similarities between the individuals in a group of rabbits are sufficiently strong for us to be able to place them all in the species called 'rabbit'. This is now trivially obvious, but it was not always the case. The classification of species was a major achievement of the eighteenth-century Swedish scientist Carl Linnaeus.

Networks, too, can be classified into various types. And the way the classification is done is in terms of features of their deep, mathematical structure. I can offer immediate reassurance that there is no maths in this chapter, or indeed in the entire book, just descriptions of the features of networks which matter to us.

There are numerous types of network. But most of those which have been found in the social and economic worlds fall, rather fortunately, into one of just three different categories. When we are looking for real-life networks, we do not have to worry about whether an actual network is one of a hundred different potential types, but most of the time only whether it is one of the three pervasive network 'species'.

And from the perspective of the policy maker, each of these three network types requires a different approach, a different strategy in terms of trying either to generate or to stop a cascade of behavioural change, of positive linking, across the agents in the network. This does not solve the problem of the uncertainties facing policy makers in a networked world – the real world of the twenty-first century. But it helps to reduce the dimension of the problem they face. If we can obtain both a good indication that network effects are important, and also a reasonable approximation to the structure of the relevant network in any given context, we can identify its type and offer guidance on what kinds of strategies are most likely to work.

What, then, are these types? Simon's *Biometrika* concept is in fact the fundamental basis for one of the three varieties of network, and it is of considerable practical importance in human affairs. In its modern guise, this genre of network is known as 'scale-free', for reasons which need not detain us.* It has a particular mathematical structure which is of great interest to natural scientists and especially to physicists.

In a scale-free network most agents have only a small number of connections to others, whilst in turn a small number of agents are connected to very many others. There is a precise mathematical relationship which describes, in an idealised scale-free network, the probability of an agent having a given number of connections, and how this probability declines as the number of connections rises. Lots of people with a few links, a few people with lots of links. The process of preferential attachment describes how such a network evolves, and how its structure is then reinforced.

This structure is important, because it means that the highly connected agents, the 'hubs', may exercise a powerful influence on the behaviour of other agents on the network. This could, of course, be because they are important people of stature, power, prestige. They exercise potential influence over others and have

* The Wikipedia entry is helpful on this topic.

so many connections precisely because of their social standing.

A simple, stylised example of a scale-free network is plotted in Figure 6.2. Let me say straightaway that there are subtle nuances across different kinds of scale-free networks, but this is a complication which we can leave aside for the moment. The chart is not intended to represent any actual network, but to give a simple portrait of what a scale-free network looks like.

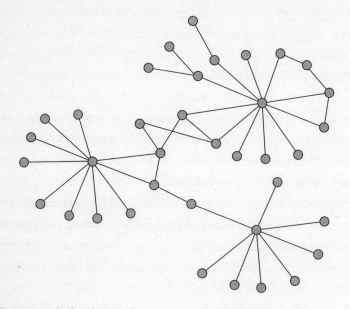

Figure 6.2 Stylised representation of a scale-free network

The circles represent the agents, whoever or whatever they may be, and the lines indicate whether a pair of agents is connected. We can easily see in this chart three 'hubs', which have lots of connections, and in contrast most other agents in the network are linked to only a small number of others, in many cases they have just a single link. Again, for emphasis, this is not meant to be a real-life network of an actual situation, it illustrates the principles.

We have come across an example of a scale-free network in

the England of the 1550s. A scale-free network of latent, potential influence was already in place. Many people would have heard of the Archbishop of Canterbury, the head of the Church of England, and other prominent clerics, bishops, preachers would also have been well known, some on a regional rather than a national scale, but famous nevertheless. The English martyrs took the gamble that they could convert, as it were, this latent network, one in which many people had heard of them, into a network which really did influence the opinions and behaviour of others.

In such circumstances, we speak of a 'weighted' network, in which a person who is well known carries more influence over someone to whom he or she is connected than an individual with few connections. But a key feature of scale-free networks is that the hubs are important even when the network is unweighted. In other words, even when an individual does not have any special power or prestige but is nevertheless well connected, he or she can exercise a strong influence on outcomes simply because of their sheer number of connections. Even if the agent has only a small probability of altering the behaviour, choice or opinion of each of the other agents to which it is connected, its influence may be strong simply because it has so many connections. A small number – the probability of influencing any given agent – multiplied by a large one – the number of connections – is still pretty big. And, crucially, in relative terms it is much bigger than two small numbers multiplied together.

Gene Stanley, a professor of physics at Boston University and editor of *Physica A*, gave a practical illustration of the influence of hubs in unweighted scale-free networks when he and colleagues discovered in 1996 that the distribution of the number of sexual partners across a sample of individuals essentially had this structure. Most people had relatively few, and a small number had very many indeed. These latter were in other respects perfectly ordinary individuals, without the prestige of, say, the Archbishop of Canterbury or even, more plausibly in this context, a film star or a famous rock singer. They were part of an unweighted network.

The implications for public health are quite disturbing. If just one of the hubs in any particular social network is carrying a sexually transmitted disease, the best way to contain the spread on the network is to identify and cure that particular hub. We can test and, if necessary, cure large numbers of less well-connected agents, but if that hub is still out there, our efforts will be in vain. Successful policy in this context requires very precise identification and targeting of a small number of individuals.

In practice, of course, we do not observe the perfect Platonic idea of a scale-free network. But, as ever, the question is whether empirically observed networks are sufficiently similar to warrant the description 'scale-free'. The formal mathematics of this gets pretty hard pretty quickly, but we can safely put to one side the various subtleties involved. Broadly speaking, however, there is a range of social and economic networks which can be approximated by this concept. This idea is already of considerable practical significance. Malcolm Gladwell's book *The Tipping Point*, published in 2000, has deservedly been a huge best-seller. Gladwell conveyed to a wider public the essential idea that, in many contexts, a relatively small number of people can exercise a decisive influence on the eventual outcome.

Marketing departments of global consumer goods companies have seized upon the idea, and very substantial amounts of money are spent on trying to identify the small number of people in any particular market who are believed to exercise undue influence on whether or not a brand or product is successful. They are making deliberate use of the concept of positive linking. If they can target and persuade the hubs, their product is likely to be very successful. As we shall soon see, the world is not always like this, and much of this expenditure may very well be completely wasted.

Gladwell identified three key ways in which individuals might be important in any particular social or economic network. He termed one group of such individuals 'mavens'. These are the people others rely upon to provide them with new information. Gladwell argues that mavens are crucial to behaviour spreading across a network by word of mouth, due to their knowledge and skills and the fact that other people regard them as having such qualities. Allied to these are the 'salesmen', who are subtly different in that they are uncannily influential in persuading others to adopt their particular buying habits or opinions.

We can usefully regard such agents as being the hubs in a scale-free network, if indeed the relevant network is of this particular kind. Gladwell's third set of potential influencers are rather more general, those he called 'connectors'. These are people with a 'truly extraordinary knack of making friends and acquaintances'. So at its most basic, they have many connections, exactly like the hubs in a scale-free network.

But the concept blurs into an entirely different one, a type of network where 'hubs' as such do not exist. A popular rendition of this is the Kevin Bacon 'six degrees of separation', the basis of a popular trivia game in which players are required to calculate the shortest number of links between two separate actors, or indeed any two individuals. The basic idea is that suppose A knows B and B knows C but C does not know A, nevertheless C is connected to A by only one degree of separation. So if B is persuaded by C to change his or her behaviour or opinions, then there is a much higher chance that A will also change than if A were not linked to B at all.

The idea that individuals can be connected in at most six

degrees of separation is an extraordinarily powerful one. A colleague of mine was recently reading a biography of Sir Roger Casement. Casement was born into a military Protestant family in Dublin, and became a quintessential member of the British establishment. However, his longstanding interest in what we now term human rights led him first of all into opposition to existing colonial empires, and eventually into active membership of the Irish Republican Army, at a time when what is now the Republic of Ireland was part of the United Kingdom. Casement tried to get German support for what proved to be the abortive republican uprising in Dublin at Easter 1916. However, Britain happened to be fighting the First World War with Germany, so Casement was tried and executed for treason.

Now, I have only three degrees of separation from Casement. My grandfather, sadly long dead now, often drank as a young man in the Jolly Gardeners public house in Rochdale, which was the preferred pub of one Jack Ellis. Although Ellis was considerably older, in the tight local community of the time, almost everyone in it knew everyone else. Ellis was a barber but also the public hangman in Britain. Casement was executed almost one hundred years ago by Ellis.*

This may seem a freak, a curiosity, but such strange connections are entirely typical. J. V. Stalin requires much less introduction than Roger Casement. With Stalin, my degree of separation is four. In the 1990s, my wife spent some time at a research institute in Stockholm. The father of one of her colleagues had been a prominent member of the Social Democrat party in Sweden from the 1930s, when that party was starting its very long reign in power in that country. However, he was a political defector,

* Incidentally, his immediate successor, by pure coincidence, came from the neighbouring town of Oldham, where he ran, without any apparently intended irony, a pub called Help the Poor Struggler.

having been general secretary of the Swedish Communist Party in the 1920s. In that role, he had visited Moscow several times, where he met not only Stalin but doomed Politburo colleagues of his such as Bukharin and Kamenev, executed in the 1930s purges. The Swede only escaped the same fate himself by seeing the light and becoming a social democrat, for a very large proportion of foreign Communists visiting Moscow in the 1930s finished up either in the execution cellars or being worked to death in the labour camps.

*

This feature of scale-free networks, that most agents in the network are typically linked by only a small series of connections, is shared by another type of network which is often found in social and economic situations. But the implications of the network for policy are fundamentally different. This is the 'small-world' network, discovered by Duncan Watts and Steve Strogatz of Cornell University in 1998. Their paper, entitled 'Collective Dynamics of Small World Networks', was published in the leading scientific journal *Nature*, and its importance is shown by the fact that over the next decade it proved to be the single most cited article of the thousands which were written on the new and rapidly expanding science of networks.

The many subtleties of the differences and similarities between small-world and scale-free networks need not concern us here, and there is a large technical, mathematical literature on them for anyone interested. But as noted already, they have the property that moving around the network, as it were, is usually pretty easy. Not many links need to be accessed for any pair of agents to be connected. Yet the two types of network are at the same time profoundly different.

Scale-free networks are distinguished by a small number of highly connected, potentially highly influential, individuals. There are no such people in a small-world network. Most individuals tend to have relatively small numbers of connections, which do not vary a great deal across the population. In other words, in a scale-free system people are characterised by the huge differences between the numbers of connections they have. In a small world, the striking feature is not the differences but the similarities.

The basic social structure of the two is radically different. Small-world systems are essentially overlapping groups of 'friends of friends'. So if X knows Y and Y knows Z, we have seen that X has the potential to influence Z's behaviour through their mutual contact Y. But in a small-world network, the contact may not need to go through Y at all. For the chances that X and Z also know each other, that they are directly connected, is high. It is by no means a certainty, but two agents who share a mutual connection are themselves likely to be directly connected.

So at a local level, when we zoom in on any particular part of it, this type of network will look pretty dense, with lots of direct connections between the agents under our focus. How can things travel across this sort of architecture, this sort of structure? To get to another part of the network altogether, it seems that many links might be required to be traversed. We have to go through successive groups of friends of friends to reach distant parts of the system. And by the time we have got there, the chances of exercising influence on the behaviour of people will be very small. So many connections have to be brought into play, and at each stage there is always the chance that our friend might just say 'no', might just persist with his or her previous choice, his or her existing opinion, rather than be influenced by us.

The feature which brings the agents much closer together, which enables this type of network to be described as a *small* world, is that a few people will have long-range connections across different groups. An individual might be an active member of a local running club, and at the same time be involved with a more sedentary hobby group, such as stamp collectors or model railway enthusiasts. Few people will be members of both. But the ones who are have the potential to spread ideas, choices, opinions, across disparate groups, sections of the network which appear, and indeed are, in other respects quite distant from each other. The word 'potential' must be emphasised. It does not mean that these 'long-range connectors' will automatically persuade people to adopt different opinions, to make different choices. In fact, they need be no more or no less persuasive than the typical agent in the population as a whole. But they have the potential to transmit behaviour across distant sections of the network.

Figure 6.3 shows a stylised small-world network with its local, overlapping connections and a few 'long-distance' connections. Again, this is not meant to be a representation of any actually existing network, but a simplified version to illustrate the key features.

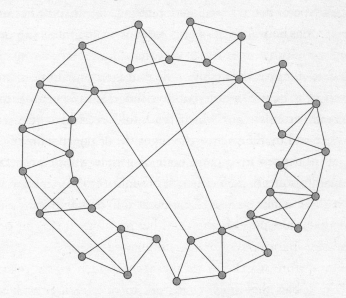

Figure 6.3 Stylised representation of a small-world network

An example of the small-world concept, where no single agent plays the role of the highly connected hubs in scale-free networks, is the network which is at work in the final chapter of Jennifer Egan's 2011 Pulitzer Prize-winning novel *A Visit from the Goon Squad*. The book is set in the future of a New York which differs in some key respects from the present-day city. Not only is energy running low, but it is socially unacceptable to use your social networks to make money. An ageing rock star is giving a concert. The promoter hires a well-connected person – a hub in network jargon – to pass on the details of the concert to his friends. He is somewhat ashamed of agreeing to do this and does not even tell his wife. Instead, he recruits a couple of much less well-connected sub-agents. The hub person is astounded when his wife suggests they go to the concert, which she has heard about from two or three people, who have all suggested it will be an exciting and novel event. Tens of thousands of people turn up and the concert is a wild success.

Clearly, an effective strategy for trying to influence networks similar to the ideal small-world form is to identify long-range connectors. But this is easier said than done. At least in some scale-free contexts, the highly connected agents might be identifiable. It is, of course, a different task altogether to persuade these to alter their behaviour, but if this could be done, there is the prospect of a cascade across the network, of positive linking. In a small world, the long-range connectors have subtle properties, not readily identified by either the number of their connections or their distinctive powers of persuasion. In fact, they look pretty much like everyone else.

*

Both scale-free and small-world networks have a considerable degree of structure which in each case corresponds to recognisable social situations. One in which there are a few highly connected agents, and the other which has an 'overlapping friends of friends' structure. But there is a third network category of practical importance which has none at all. Indeed, this lack of structure is reflected in the name: random networks. Such networks began to be investigated by mathematicians such as Paul Erdös in the late 1950s and 1960s. Erdös is well-known in mathematical circles for the fact that other mathematicians boast when they have a low 'Erdös number', the mathematicians' equivalent of degrees of separation. He was so enormously productive and collaborated with so many of his colleagues that many have an Erdös number of one.

It may seem strange that, in human social and economic contexts, the concept of random connections makes any sense at all. But it does. We have already mentioned the example of the common cold. You are travelling to work on the subway. The person next to you sneezes. Unfortunately, his cold is at the infectious stage. You have never met your temporary travelling companion, have no idea who he is, and may very well never see him again. But you have caught his cold.

Epidemiologists have analysed the spread of diseases for many years. Their basic model, developed as long ago as the 1920s, is so often used that it has its own name, the SIR model. The relevant population at any point in time can be allocated into one of three groups. Those who are susceptible to a particular disease, but have not yet caught it are in category S, those who are infected are Is and those who have been infected but are no longer so are the Rs. With infections that do not confer immunity, such as the common cold, R stands for 'recovered'. Because Rs are susceptible to reinfection, they can move back to the S category. In other cases, recovered individuals will almost certainly be immune to reinfection, so that R here stands for 'removed', a description which also applies more literally to altogether more sinister infections such as the Black Death.

So at any point in time, there is a proportion of the population which is susceptible, a proportion infected, and a proportion which has recovered. The mathematical models used in epidemiology describe how these proportions change over time. At its simplest, any individual is assumed to have the same chance of contracting the disease upon coming into contact with an infected person. So the chances of any individual flowing, as it were, from S to I depends upon this and the number of infected people he or she comes into contact with. Equally, everyone who is infected is assumed to have the same chance of recovery in any given period, of leaving the I category and moving into the R.

The above paragraph actually describes a system of differential equations. To be more precise, a system of non-linear differential equations. As these things go, it is just about as simple as you can get. But even so, it has proved to be an extraordinarily powerful scientific tool. Of course, there are many much more complicated variants, but the basic principles remain the same.

It also contains a hidden network. Can you see it? It is the contacts between an S and members of the I group. These links constitute the relevant network. No particular social structure is assumed. In fact, there is none. Susceptibles just encounter infected people at random. Purely chance encounters, like sitting next to a stranger with a cold, are the basic way in which infections spread in the fundamental model of the whole science of epidemiology.

But how does this relate to wider social and economic issues? Fashion – indeed, many aspects of popular culture – might reasonably be regarded as spreading (or not, as the case may be) across a network in which individuals are connected purely at random. Of course, to state yet again, all scientific theories are approximations to reality, and whether they are any good depends to a large extent on how reasonable their approximations are. So, yes, other factors, other types of network might impinge on the story, but an awful lot about the spread of fashion can be explained by random networks. The blurb for a contemporary fashion site called Fashion Indie makes this admirably clear: 'Fashion Indie is a collection of all things related to fashion, from news and notes on models and designers to a section called "random cool shit" that highlights, well, pictures of random stuff that's awesome.' Random stuff that's awesome!

The SIR model discussed above has been developed in more modern, networked-based versions. Individuals can be allowed to have different degrees of susceptibility, but the basic principle of a random social network is retained. From the longstanding work of epidemiologists, we know a key feature of such networks. Specifically, we know the things we need to know – the 'knowable knowns' – in order to work out whether an infection has the potential to spread across the network as a whole. From a policy perspective, this tells us the specific goals we have to achieve if we want an idea or mode of behaviour to percolate across a random network (or, conversely, what we need to do to prevent something spreading).

We have already encountered the idea that social and economic networks are 'robust yet fragile'. In other words, most of the time, most new events, shocks, new ideas, new choices do not percolate very far across them. But, occasionally, one event which is otherwise indistinguishable from those that fail encounters success. The network proves 'fragile', and the small initial change spreads through it. In random networks, we know that there is a distinct boundary between fragile and robust. Unless something gets enough traction within the network, enough people either catch or adopt it, it has literally zero chance of spreading on a substantial scale. Indeed, it will simply fade away and disappear altogether. Calculating where a particular infection is in relation to this boundary might be difficult in practice, but we know how to do it. The precise calculation will vary from context to context, but in principle we know how to calculate it.*

So, if we are faced with a random network and want to encourage the spread of a product or a mode of behaviour, we basically have to get enough people to buy into it to get over a critical mass, otherwise our efforts are wasted. Small-world networks are more complicated, though they often resemble random ones in this respect. The same is not true of scale-free networks. Although the probability of a large cascade may be very small, even if just a few people adopt a new fashion, a new behaviour, a new idea, there is still a chance that it will spread across the network as a whole. Obviously, it would be desirable to persuade more people rather than fewer in the first wave of marketing, but even if it does not go well, we can still cross our fingers and hope. All is not lost. It almost certainly is, but we might still succeed.

Here, then, is a way in which network structure might influence the strategy of someone wanting to influence the whole network in some way, including a policy maker or a marketing department. Facing a random network, or one which is more similar to this than any other kind, we just have to have a blitz, to try to persuade enough people so our offer leaps over the critical mass. Strategies on other networks might have to be more subtle. In other words, where networks are important, generalised statements about policy effectiveness across different policy domains lack validity because of the different natures of the different types of network which predominate in the social and economic worlds. The effectiveness of a policy will be contingent on the type of network upon which it is being enacted.

* A clear illustration of this in the basic SIR model is given at: http://en.wikipedia.org/wiki/Compartmental_models_in_epidemiology

Inspired by Malcolm Gladwell, many marketing departments have strategies that concentrate on finding the hubs of the network, the 'influentials' who are believed to be the key to success. The concept has many followers. A Google search on 'influentials' comes up with 220,000 sites. The technical phrase 'discovering influentials viral marketing' yields no fewer than 55,000. Viral marketing is a relatively new concept that takes its name from the spread of viruses in epidemiology. How can we get awareness of our brand, and then hopefully the sales of it, to percolate, to cascade across pre-existing social networks? Word of mouth is one answer, text messages another, and web-based tools include techniques such as video clips and interactive Flash games.

The idea of influentials has a much broader scientific base, and essentially goes back to the work of Paul Lazarsfeld in the 1940s. Born in Vienna in 1901, Lazarsfeld was one of the many Jewish people welcomed by America in the 1930s and who did so much to enrich intellectual life in their new country. He founded Columbia's Bureau of Applied Social Research, and was a towering figure in twentieth-century sociology in America.

The 1940s was the decade when the mass media which dominated the second half of the twentieth century really came into effect. Newspapers had existed for many years, had been more recently augmented by radio and were now being massively enhanced by the new medium of television. For an internet-oriented population, it is hard to remember and grasp the truly dramatic impact of television. For the first time in human history, millions of people could not just listen to but actually see those who were trying to entertain, amuse or persuade them.

The latter possibility, that the media might influence people's choices and behaviour, soon became a matter of serious concern.

Vance Packard published a stupendous best-seller entitled the *Hidden Persuaders*, allegedly exposing how advertisers and politicians manipulated public opinion subliminally through the mass media. Such worries persist to this day. In Britain we had in the spring of 2011 one of our very rare referendums. This was on whether to change the first-past-the-post system of voting, which has existed in Britain since elections began, in favour of some newfangled alternative. The change was rejected decisively by a majority of more than two to one. Yet this did not deter vociferous denunciations of the impact of the 'Murdoch press' on the result.* The electorate, it was claimed, were prevented from having a 'proper conversation' on the matter by the wicked media. In America, criticisms of the influence of both the liberal press and right-wing radio talk-show hosts are a routine feature of the broader political discourse.

Lazarsfeld articulated a sophisticated theory of how such influence might be exercised. He proposed a concept known as the 'two-step flow of communication'. On this hypothesis, individuals were not so much influenced directly by what they read, heard or saw in the mass media. Rather, the flow was mediated by influentials. Ideas, choices, behaviour go first from the media to influential individuals, who in turn then spread their own interpretation across the population as a whole. These influentials need not occupy prominent positions or be in any way obviously remarkable. Indeed, they are very similar to their peers in terms of interests and socio-economic characteristics. In the context of modern network theory, we have seen that the hubs of the network, the agents with large numbers of connections, may indeed exercise strong influence simply because they have so many connections.

* In the run-up to the referendum, the several right-of-centre UK newspapers owned by Rupert Murdoch all vigorously opposed the proposed change in their opinion columns.

They do not need to be especially knowledgeable or persuasive; they just know a lot of people.

The theory was seen as a way of explaining why some advertising campaigns, some political messages, succeeded whilst others failed. If, as was asserted to be the case, the mass media and the advertising budgets of giant firms now controlled people's choices, the fact that many such campaigns failed needed to be explained away. And the two-step theory formed the basis for this rationalisation. It wasn't so much that the influence of the media was uncertain, but the fact that the influentials it influenced might in some cases distort the message in ways that would render it less palatable. As a result, the concept, argument or product involved would not be widely adopted.

Like any seminal concept, the two-step flow of communication theory has generated a large literature, some of which puts forward supporting evidence, some of which is critical. It is useful to reflect, however, on Lazarsfeld's original article, a study carried out with Bernard Berelson, and Hazel Gaudet. This focused on how people decided how to vote during the 1944 US Presidential campaign. The research team came to the project with the hypothesis that the media would exercise a direct influence on the views and decisions of the electorate. To their surprise, informal, personal contacts were mentioned far more frequently as the source of influence on voting behaviour than exposure to the mass media. It seems to have been more or less axiomatic that the mass media did in principle exercise a powerful influence. So the direct contradiction of this hypothesis by the empirical evidence led to a theory which retained the concept of media power, but which at the same time accounted for the fact that it might not always be influential.

Echoes here of rational agent economic theory, but at least Lazarsfeld saw the need to confront his theory with empirical evidence. When the evidence failed to support the theory, just as the Ptolemaic epicycles could always account for observed deviations from the theory that the sun went round the earth by adding another complication to the model, a pretty spectacular addition was made, namely two-step communication flow.

Lazarsfeld went on to develop and examine this theory further with Elihu Katz, who published his reflections in the *International Journal of Public Opinion Research* in 2001. Katz noted that most of the work in the field confirmed Lazarsfeld's original finding that the media exercised mild influence on public opinion, but that interpersonal transmission was in general much more important.

For many sociologists and media-studies academics, the idea that the mass media lack this power is unacceptable, regardless of the fact that the empirical evidence points towards it. As Katz points out, one of the attacks is based on the view that the media generate a false consciousness of reality amongst the broad masses. How can it possibly be that people in general do not subscribe to the views of the liberal elite, which are so self-evidently correct? They must obviously have been brainwashed.

Despite the evidence, the idea that a small number of individuals are responsible for the spread of ideas, opinions, choices, behaviour, has become very widespread. Even by the mid-1990s, the two-step theory of communication and the concept of influentials had generated more than 4,000 academic articles, and the number has grown dramatically since then. The diffusion of technological innovations, communications studies, marketing – all these areas of study were affected.

*

Network theory moves us completely beyond these concepts. The deep insight that social and economic networks have the property of being robust yet fragile tells us exactly why some campaigns, whether commercial or political, fail to spread, and why others have a dramatic impact. It is an uncomfortable insight. The element of chance and contingency is prominent in the story. We have *inherently* less control over situations in which network effects are important than we would like, no matter how clever we might be in trying to design policies to bring about desired outcomes.

We have also seen that if the network in any particular context is best approximated by a random or a small-world one, then the very concept of influentials is not terribly helpful. Of course, if the relevant network is similar to the scale-free template, there are certainly individuals who do have the capacity to exercise a strong influence. These hubs, even when they are otherwise ordinary, unremarkable people, are important simply because they have so many connections. Even if each of their links has only a small probability of being influenced by them, the fact that they have a large number of connections means they have a good chance of altering someone else's behaviour. And if the individuals themselves are seen as being important, figures of prestige, then their impact is doubly important.

Indeed, in this particular context, when we are operating with a weighted scale-free network, the role of influentials and the media in forming public opinion takes on an entirely different meaning from the one articulated by Lazarsfeld and Katz. Friedrich Hayek, in his 1949 essay 'The Intellectuals and Socialism', considered the question of how ideas of planning and socialism had come to a dominant position in the market-oriented economies of the West. He attributed this to the role of intellectuals. By the word 'intellectual' he did not mean an original thinker. Rather, for Hayek, intellectuals were 'professional second-hand dealers in ideas', such as journalists, commentators, teachers, lecturers, artists or cartoonists.

The opening paragraph of Hayek's essay reads:

In all democratic countries, a strong belief prevails that the influence of the intellectuals on politics is negligible. This is no doubt true of the power of intellectuals to make their peculiar opinions of the moment influence decisions, of the extent to which they can sway the popular vote on questions on which they differ from the current views of the masses. Yet over somewhat longer periods they have probably never exercised so great an influence as they do today in those countries. This power they wield by shaping public opinion.

During the second half of the twentieth century, the influence of the concept of socialism in the West weakened. The view that governments can plan, predict and control outcomes still pervades much political thinking, and the role of the state in the economy remains very much larger than it was in the first half of the twentieth century. But market-oriented thinking has become much more powerful. In part, this is due to a sustained effort within economics to create a powerful intellectual argument for this view, and in part due to the disastrous performances of planned economies, whether in the former Soviet bloc, Africa or elsewhere. Acute poverty is now a distinguishing feature of anti-capitalist regimes rather than the more market-oriented ones. However, despite the overwhelming evidence of the success of the market-oriented economies of the West based upon capitalist principles, it still attracts opposition, often virulent, from many intellectuals. The system is indeed prone to occasional crises, such as the financial one of 2008–9, but it has brought unprecedented prosperity to literally billions of people across the world.

The success of liberal intellectuals – using the word 'liberal' in the American sense – is much more complete in social and cultural policy and debate, a phenomenon which arose mainly during the final quarter of the twentieth century. Hayek had earlier noted that 'it is perhaps the most characteristic feature of the intellectual that he judges new ideas not by their specific merits but by the readiness with which they fit into his general conception, into the picture of the world which he regards as modern or advanced'. Just one example will suffice. The view that all types of family structure are equally valid has become an article of faith with the metropolitan liberal elite. Yet it is rejected decisively by all serious studies of the problem. Children brought up in households of never-married single mothers, for example, face a very much higher probability of being consigned to a life of poverty and crime than in stable two-parent households.

Hayek himself is very clear on how such a small minority can set so decisively the terms of debate on social and cultural matters. He writes: '[Socialists] have always directed their main effort towards gaining the support of this "elite" [of intellectuals], while the more conservative groups have acted, regularly but unsuccessfully, on a more naïve view of mass democracy and have usually vainly tried to reach and persuade the individual voter.' He goes on later in the essay to state, 'It is not an exaggeration to say that, once the more active part of the intellectuals has been converted to a set of beliefs, the process by which these become generally accepted is almost automatic and irreversible.'

Hayek did not articulate a formal model of how this process operated, for network theory scarcely existed at that time, and was in any event mainly confined to the abstract musings of high mathematics. But he was essentially articulating a quite different process from that of Lazarsfeld and Katz. First of all, the small minority of influentials – Hayek's intellectuals – are targeted. If successful, they then define the terms of the debate and influence the media. Initially, their impact is on the more elite outlets, which in turn affect how the media as a whole frame the arguments. So the intellectuals affect and determine mass opinion both themselves directly, through appearing in the media, and indirectly, by influencing the arguments and how they are presented.

Simple network models based on the principle of binary choice with externalities which we saw in the previous chapter illustrate clearly how Hayek's idea can work in practice. As before, we have a population of agents who face a choice between two alternatives, and each person has his or her own degree of persuadability. An agent switches behaviour if the proportion of all the others to which it pays attention – is connected, in other words – is above its persuadability threshold. Last time, this was a simple matter of counting heads. But this time round, each agent to which an individual is connected is weighted by the number of agents to which it is itself connected. People with lots of connections, known by many, have prestige in this version of the model, and the weight given to their choice reflects this.

So, putting the agents on a scale-free network, we can start off with, say, 20 per cent of the population holding liberal social views and 80 per cent conservative. If these same proportions are reflected in the hubs, the key influencers in the network, when we run the model many times, the proportion which ends up as liberal on average falls. Indeed, in more than half of the solutions, the proportion which is liberal eventually falls below 5 per cent. But suppose now that we start off with exactly the same overall proportion, but specify that the most connected 2 per cent of the population, a tiny minority, start as liberals. It is then conservatism which ends up close to elimination most of the time. By making this initial percentage smaller we can of course get more mixed results. But the message is clear. In situations where highly connected people carry special influence because of the number of their connections, they can indeed be as decisive as Hayek intuitively believed them to be.

Networks are clearly very important in determining the outcomes of many social and economic processes. As we have seen,
they introduce new levels of uncertainties for policy makers, a
feature not just of network models themselves but, far more significantly, of the real world. We have begun in this chapter the
task of trying to scale down these uncertainties. Scaling them
down by finding ways which will not only help us to understand
whether network effects are present in any particular situation,
but which offer practical advice to policy makers depending on
the particular type of network which is relevant. The theme of
policy making in a networked world is one to which we return at
greater length in the final chapter.

A disturbing feature of many of the examples which have been
given is that, when network effects are present, it is the networks
which are more important in determining outcomes than the
objective attributes of the various choices and courses of action
on offer. This is virtually a complete reversal of the principles of
the rational model of behaviour according to economists. There,
the qualities and features of the alternatives are crucial to their
success or failure. If our world is like this, the phrase attributed
to Ralph Waldo Emerson has distinct resonance: 'Build a better mousetrap and the world will beat a path to your door'. In a
world where network effects are important, this appears to be by
no means guaranteed. The mathematics of the network models
suggest that almost any brand of mousetrap could end up being
the market leader, with more and more people beating a path to
the door of the producer, almost without regard to its efficacy if
it is fortunate enough to be the recipient of the benefits of positive linking.

The maths certainly gives these results. But are we simply falling into the trap of mainstream economists, of reifying abstract
mathematical approaches, albeit of a completely different, much

more modern variety than the ones which economists use? Maths is all very well, but what evidence, what actual human behaviour, can be shown to be compatible not only with how these models work, but with their results, which seem to defy, if not common sense, then at least conventional wisdom? These questions are the focus of the next chapter.

7

Copying Is the Best Policy

Herb Simon's revolutionary insight for economics was that, in most situations, agents are unable, literally unable, ever to compute the optimal decision. It is not a question of placing constraints on their behaviour, such as the costs of gathering and processing information, and then assuming that they make the best possible decision in the light of these constraints. It is that they can never know what the best decision is, not just before but even after the event.

In Simon's theory, agents are still evaluating alternatives, but they decide amongst them on the basis of the features, the attributes, of the choices they consider. The model of behaviour outlined in the previous chapter takes us even further away from the model of rational choice in economics. It suggests that the actual benefits offered by a product, service, idea, lifestyle choice, etc. over those of its competitors are of little relevance to the outcome. Once something starts to become a little bit more popular than its rivals, for whatever reason, a feedback mechanism is set up. Its popularity makes it even more popular. It is positively linked.

All theories are approximations to reality, a familiar mantra by now. One theory about how an agent makes decisions postulates that it obtains all the information relevant to that decision, that it is capable of interpreting and processing this information so that it makes the best choice for itself, and that it operates independently, not being influenced directly by other agents, with fixed tastes and preferences.

This is, of course, the theory of Rational Economic Man. And it was being formalised and developed for the first time during the final quarter of the nineteenth century and the early years of the twentieth. Even then, its assumptions were not completely true. But perhaps they were not too bad; reasonable enough approximations to form a working model which revealed something about how the world worked.

I have a cherished copy of the April 1910 edition of *Bradshaw's*. Very few readers will know what *Bradshaw's* is. And when I explain that it is a timetable of all the trains which ran in Britain and Ireland in that month in that year, you may feel that a lack of acquaintance with this volume may be an advantage rather than a handicap. Yet it is a fascinating social document. Railways made journeys very much faster than they had ever been. The London to Edinburgh stagecoach even with the new turnpike roads of the late eighteenth century and better technical design of the coaches, could not reduce the journey time to less than thirty-six hours. By rail in 1910, the journey could be accomplished in a mere eight. On the other hand, when connections had to be made, perhaps from the main line to a branch, railway companies thought nothing of keeping passengers waiting for two or three hours for no apparent reason. Even on today's overcrowded roads, these journeys can be accomplished more quickly by car.

But the real interest in the present context is the advertisements that appear throughout the timetable. Not surprisingly, there are details of many hotels that the upper- and wealthier middle-class readers of *Bradshaw's* in 1910 might be drawn to. (One boasts that 'staff in full livery greet the train and carry your luggage'.) Other adverts are for quite simple, straightforward products. We read that 'Keating's Powder Kills Beetles', over a graphic picture of several dead beetles and another crying into a handkerchief. Apart from the prices of the variously sized tins in which it is sold, that is the entire content of the advert. Another boasts that 'Eux-E-Sis' is a 'delightful cream for shaving, no soap, no brush, no jug', though it is also careful – perhaps in an early example of today's paranoia that inadequate instruction might result in litigation – to specify that 'a razor is needed'. One of the most sophisticated products on offer is Benson's gold watch bracelets, which are described as containing 'lever movements' and to be 'warranted timekeepers'.

In short, the adverts essentially described what the products did. And they had limited functions which would be easy to evaluate. There was no promise of a fulfilled lifestyle, of sexual ecstasy, of the whole gamut of human emotions which twenty-first-century advertising evokes. We are simply told that Keating's Powder kills beetles and that Benson's watches tell the time.

*

Branded products were relatively new in Edwardian Britain. Of course, what we can think of as brands had existed almost as long as humanity. At a time when most people were illiterate, the Red Lion pub might have a sign outside painted with an image of that mythical beast. The sign both informed potential consumers that alcohol could be obtained there, and served as an indicator of the quality and reliability of the offer. In the great fourteenth-century allegorical poem *Piers Plowman*, in which the Deadly Sins are personified, Glutton is on his way to church to confess. He is waylaid by none other than Betty the Brewster herself, standing outside her premises delivering personalised messages to her potential customers, almost in the manner of recommender emails in our own times. She targets Glutton for his fondness for 'hotte spyces', and he succumbs, imbibing no less than a 'galloun and a gyll' (about five litres) of strong ale.

Products designed to be sold to a mass market were essentially an innovative feature of the nineteenth century and in particular the final quarter of that century. The Industrial Revolution ushered in for the first time a social and economic system in which sustained long-term growth was a fundamental feature. Ironically, it was none other than Karl Marx who was the first economist to understand that this was a permanent feature of the new capitalist system. Times were hard for the emerging industrial working class, with punishing hours and meagre wages. But gradually the benefits of growth seeped through. Millions of people began to have that little bit extra to spend over and above what was required for mere subsistence. Branded products were developed, designed to be sold to a new mass market, and corporate structures and technologies evolved to enable them to be produced and delivered.

Alongside them, the new profession of advertising sprang up. N. W. Ayer and Son opened in Philadelphia in 1869, and were the first company to perform what is now the central offer of advertising: the design of promotional messages for clients. In common with other agencies, they brokered the rates for advertising space in newspapers, but Ayer's were the first to take charge of the content. From these modest beginnings, advertising now operates on a massive scale. Estimates of the current global size of the industry vary, but its turnover is not far short of a trillion dollars. Initially, consumer choice and awareness remained relatively limited. Mass markets were in their infancy. Moreover, the products themselves tended to be uncomplicated, their qualities both few and obvious and readily evaluated. Fashions, of course, existed, as they have done throughout human history. But awareness of the choices made by other people and of the detailed events in the wider world was, by today's standards, limited. Things had moved on since 1840, when the inhabitants of the remote Atlantic archipelago of St Kilda, the farthest-flung part of the British Isles, were still offering prayers to the health of His Majesty King William IV three years after his death. But the instant access to what others are doing and thinking, which is now taken for granted, simply did not exist.

*

So perhaps a century ago, the assumptions of the rational choice theory of economics might not have been – how shall I put it? – completely unreasonable ones to make. This is not to say that they were necessarily good ones, just that they described a theoretical world which at least bore some tenuous connection to reality. But what does the world of the twenty-first century look like?

First of all, and most obviously, the speed and scale of communication of the opinions and decisions of others changed dramatically during the twentieth century. And in particular, the internet over the most recent ten to fifteen years has revolutionised these features of the world. The technological innovation of the internet certainly makes feasible an entirely different model of behavioural decision making, in which people take into account directly the choices and opinions of others.

This development was foreseen in the remarkable work of Marshall McLuhan, a Canadian Professor of English and a communications theorist based at the University of Toronto in the 1950s and 1960s. He became well known to the general public through his 1967 book *The Medium is the Massage*. The word 'massage' is correct, though the phrase was subsequently adapted to 'the medium is the message'. McLuhan believed that each technological medium of communication 'massaged' the way in which agents viewed the world. Behaviour itself was altered by revolutions in communications technology. The medium affects society and how people behave not by its content but by its characteristics. So for McLuhan, the actual content of what was broadcast on television did not matter.

Regardless of the validity of McLuhan's theories, in a very practical sense both twentieth-century technology and now the internet have completely transformed our ability to discover the choices of others. We are faced with a vast explosion of such information compared to the world of a century ago. We also have stupendously more products available to us from which to choose. Eric Beinhocker, in his excellent book *The Origin of Wealth*, considers the number of choices available to someone in New York alone: 'The number of economic choices the average New Yorker has is staggering. The Wal-Mart near JFK Airport has over 100,000 different items in stock, there are over 200 television channels offered on cable TV, Barnes & Noble lists over 8 million titles, the local supermarket has 275 varieties of breakfast cereal, the typical department store offers 150 types of lipstick, and there are over 50,000 restaurants in New York City alone.'

He goes on to discuss stock-keeping units – SKUs – which are the level of brands, pack sizes and so on which retail firms themselves use in re-ordering and stocking their stores. So a particular brand of beer, say, might be available in a single tin, a single bottle, both in various sizes, or it might be offered in a pack of six or twelve. Each of these offers is an SKU. Beinhocker states, 'The number of SKUs in the New Yorker's economy is not precisely known, but using a variety of data sources, I very roughly estimate that it is on the order of tens of billions.' Tens of billions! Tens of billions of alternatives from which a customer can choose.

So, compared to the world of 1900, the early twenty-first

century has seen quantum leaps in both the accessibility of the behaviour, actions and opinions of others, and in the number of choices available. Either of these developments would be sufficient on its own to invalidate the economist's concept of 'rational' behaviour. The assumptions of the theory bear no resemblance to the world they purport to describe. But the discrepancy between theory and reality goes even further.

Many of the products available in the twenty-first century are highly sophisticated, and are hard to evaluate even when information on their qualities is provided. Mobile (or cell) phones have rapidly become an established, very widely used technology (despite the inability of different branches of the English language to agree on what they should be called). Google searches on 'cell phone choices' and 'mobile phone choices' reveal, respectively, 34,300,000 and 27,200,000 sites from which to make your choice. And how many people can honestly say they have any more than a rough idea of the maze of alternative tariffs which are available on these phones?

So here we have a dramatic contrast between the consumer worlds of the late nineteenth and early twenty-first centuries. Branded products and mass markets exist in both, but in one the range of choice is rather limited. In the other, a stupendous cornucopia is presented, far beyond the wildest dreams of even the most utopian social reformers of a century ago. An enormous gulf separates the complicated nature of many modern offers from the more straightforward consumer products of a mere hundred years ago. And, of course, we are now far more aware of the opinions and choices of others.

*

There is very strong evidence from the discipline of psychology that, in a world such as this, the assumptions made by the rational choice theory of economics do not make much sense. William Hick was a British academic psychologist who spent most of his working life at Cambridge. In 1952 he published a paper in the *Quarterly Journal of Experimental Psychology* entitled 'On the Rate of Gain in Information'. It made him famous, at least within the world of psychology, and its results became immortalised in Hick's Law.

Hick carried out an experiment involving semaphore lamps and their corresponding Morse code keys. (More irreverent readers may recall the Monty Python sketch in which Heathcliff and Cathy enact *Wuthering Heights* by semaphore, to be followed by Wyatt Earp and Doc Holliday in *Gunfight at the OK Corral* in Morse code. But this code, in which each letter of the alphabet is represented by a combination of short and long clicks, was widely used for open communication, not least within the military during the Second World War.) Hick had lamps lighting at random every five seconds. The reaction time to choose the corresponding key was recorded, with the number of choices ranging from two to ten lamps. Hick's Law essentially describes the time it takes for a person to make a decision amongst the possible choices he or she has. Technically, given a number of equally probable choices, the average time required to choose amongst them increases with the value of log base 2 of the number of choices.*

But what does this mean? Rather obviously, it suggests that the time taken to evaluate and choose rises with the number of choices. It also tells us that the amount of time required rises more slowly than the number of alternatives. The value of 'log base 2' of any number is simply the number of times 2 has to be multiplied by itself to get that number. So the log base 2 of 2 itself is just one, the number of times 2 is multiplied by itself to get the

---

* Plus one, to be accurate, and with an empirically determined constant multiplicative factor.

value 2. Log base 2 of 4 takes the value 2, for 4 is 2 multiplied by itself two times, and so on.

Now, the value of this formula when applied to the number 1,000 is just under 10. Hick's Law suggests that if it takes me one unit of time, whatever that might be, to decide between two alternatives, it will take me ten times as long to pick one from 1,000. Log base 2 of 27,200,000 is nearly 25. So I Google 'mobile phone choices' and get 27,200,000 sites from which to choose just to find out more about the choices open to me. The one unit of time it would take me to decide in the extremely unlikely event of only two sites popping up in a search, expands by a factor of 25 to decide amongst all the alternatives, even if in some miraculous way they could all be presented to me on the same page. And moving between pages just to find out what the choices are is itself time consuming.

Not surprisingly, most people in this situation simply do not bother even to try to look at all the alternatives. Indeed, as we have seen, on average in such searches, 98 per cent of all searchers just hit one of the top three results. And no fewer than 60 per cent select the number one. They behave in exactly the way postulated by Simon, by trying to strip down the vast dimension of the choices which faces them and use instead a simple rule of thumb: hit one of the first three sites in the list.

In so doing, as we have already seen, their choices are influenced by a network. The random network of other people who previously selected the sites which have become the three most popular relating to that particular search. Before I typed 'mobile phone choices' into my search engine, I may not even have had a preference at all for the site I would investigate in more detail, may not even have known it existed. But now I do. My preference has not just been altered, but it has been defined by the previous selections made by agents on a random network.

This is a key reason that 'copying' has become the rational way to behave, the rational way to make choices in the twenty-first century. The range and complexity of choice are so vast that the only way in which agents can cope is by adopting behavioural rules which spectacularly reduce the dimension of the scale and scope of the choices available to them. The word 'copying' is, I should stress, being used as a shorthand description of the mode of behaviour in which your choice is influenced, altered, directly by the behaviour of others. Lying behind the actions may be several different motives.

An important one relates to the problem posed by Herb Simon back in the 1950s, when he argued that humans simply cannot gather and process information on the scale required by the standard economic model of rational behaviour. He suggested that in most situations we simply cannot compute the optimal decision, so instead we rely on rules of thumb. And we recall that in the late 1930s Keynes concluded that 'we have, as a rule, only the vaguest idea of any but the most direct consequences of our acts'. Keynes's response was the same as Simon's. In such circumstances, both these great thinkers argued that we need a new definition of rational behaviour. The economic definition of 'rational' is no longer relevant. Recall Keynes's explicit statement that copying is a key element of rationality: 'Knowing that our individual judgement is worthless, we endeavour to fall back on the judgement of the rest of the world which is perhaps better informed.' So, in many situations, relying at least in part on the actions and opinions of others is entirely rational. Copying makes sense.

*

Fashion is another reason, more basic somehow, and easier to understand. It seems to be a deep-seated human instinct to like to buy, wear or choose the things which lots of other people find popular. Of course, there can be reactions in the opposite direction. Thorstein Veblen's fame rests on his 1899 classic *Theory of the Leisure Class* and the phrase 'conspicuous consumption' which has become associated with it. In economics, there is even a special class of products which go by the name 'Veblen goods'. Their characteristic is that the demand for them increases as the price rises, as a higher price signals greater status. We might, of course, see this form of behaviour as itself a fashion, albeit one confined to those with sufficient income and wealth with which to follow it. Veblen's snobs adamantly refuse to buy or adopt any form of behaviour which is remotely popular with the broad masses.

On the day I wrote the first draft of this chapter, I saw in a popular newspaper a piece on the ultra-high-heeled shoes designed and marketed, at ultra-high prices, by Christian Louboutin. The journalist points out that 'price, not taste, is the only thing that matters to the rich and famous' before going on to administer a stern warning: 'But Mr Louboutin needs to beware. If you make a brand too popular and ubiquitous, allow it to be worn by the super-tanned rather than supermodels, you are in danger of devaluing it. Remember when Daniella Westbrook was seen dressed head-to-toe in Burberry check? This British brand's credibility plummeted overnight and it took years of clever marketing, a new creative director and pushing the brand ever more crazily towards so-called "luxury" to leave its less desirable customers floundering.' Purely by coincidence, when revising the draft, I read that Michael Sorrentino, famous for appearing in MTV's pseudo-reality TV series *Jersey Shore*, has been offered money by Abercrombie and Fitch *not* to wear their clothes on the programme. Sorrentino has a lack of discernible qualifications, talent or even good looks, and indulges in behaviour which might be regarded as hedonistic or even irresponsible. The company said:

'We are deeply concerned that Mr Sorrentino's association with our brand could cause significant damage to our image.'

But fashion seems eternal. The Hittite empire reached its peak in Anatolia in the fourteenth century BC, nearly 3,500 years ago. Even here, at this immense distance of time, we can find evidence of the fashion motive at work in the decisions people made. Archaeologists have put together, in a painstaking and thorough way a database of ceramic bowls from two successive phases of occupation of Boğazköy-Hattusa, capital of the Hittite empire and the largest Bronze Age settlement in Turkey. The bowls differ in features such as size and the materials used.

James Steele and his colleagues Claudia Gatz and Anne Kandler, of University College London, published an article in the *Journal of Archaeological Science* in 2010. They asked the question: to what extent are the observed frequencies of the different types of bowl due either to pure fashion, to people simply copying designs which were already popular, or to the inherent functional characteristics of the bowls? The latter case is an example of incentives at work, of rational choice in the economic sense. Agents consciously select amongst alternatives after considering the attributes of each of the potential choices.

Using some advanced mathematical techniques, related to the discussion on preferential attachment in the previous chapter, they concluded that the pure fashion hypothesis goes a long way in being able to account for the frequency with which different types of ceramic bowls are observed in Hittite culture 3,500 years ago. However, they also found that incentives, as it were, also played a part. Rational choice, conscious selection on the basis of the features of the bowls, also had to be invoked to give a more complete understanding. But fashion explained the most.

*

So far we have invoked two motivational factors which can under-pin the theory of copying as the new rational mode of behaviour. One of these, fashion, is almost as old as humanity itself. The other is much newer, but equally fundamental in the modern world. This is the need to dramatically scale down the dimension of choice in the face of both the vast array of products and serv-ices that modern capitalism provides and the knowledge which communications technology makes available to us in terms of the decisions and opinions of others.

But there are further important behavioural principles which justify the concept of copying as the new rationality. Solomon Asch was born in Warsaw and in 1920 moved to America, where he carried out pioneering work in social psychology. His most famous experiments, like those of Hick, were done in the early 1950s. The participants were asked to compare the length of a line with other everyday objects. One such task was to compare a line with three others and say which one of the three was the same as the given line. Visually, the answer was obvious. It was not a matter of careful judgement, of close inspection and calibration. One of the three was obviously the longest.

Each group of players was seated in a room, and asked to announce their opinion out loud and in turn. In the control groups, in which everyone involved was giving their own independent opinion, almost everyone gave the correct answer. Asch then repeated the experi-ment with people planted to give an incorrect response. And these were seated so that they would be the first to speak. The 'real' par-ticipants, as opposed to Asch's people, were put at ease by carrying out two sets of comparisons in which all the plants gave the correct answer. On the third and subsequent trials, however, all the plants gave the same wrong one. To Asch's surprise, no less than three-quarters of the genuine responders repeated the incorrect answer at least once. And around a third almost always did so.

Asch then carried out modifications of the experiments. One was to discover how many plants were needed to make the real participants reply incorrectly. If there were only one or two, they were almost always ignored, but three was the critical number. Once three gave the incorrect answer, at least one real participant would follow their line.

There have been many subsequent variants of the Asch experiments. One was carried out by the French-Romanian psychologist Serge Moscovici in the 1970s. Groups of six were formed comprising four participants and two plants. They were all shown thirty-six slides, each slide a shade of blue, and asked to state the colour out loud. There were two groups in the experiment. In the first group the plants were consistent and answered green for every slide. In the second group they were inconsistent and answered green twenty-four times and blue twelve times. Moscovici found that when the plants answered consistently, the influence of the minority in inducing incorrect responses from the honest participants was distinctly stronger.

A number of criteria have been determined to judge the potential influence of a dissenting minority in more general situations, such as when discussion is permitted. Perhaps unsurprisingly, it is stronger when the minority is consistent, when it is perceived by others to be competent, and when the majority genuinely feels uncertain. Again, not surprisingly, if it is the majority which dissents from the correct answer, the impact is more powerful still. What *is* surprising is that even a minority can induce people to give a response which is obviously incorrect.

It is certainly surprising in the context of the rational agent model, where, once again, the agent is postulated to gather all information and to make the best choice, operating independently. How hard can it be to say whether lines are longer or shorter than another when the answer is clear from simple inspection? Well, it certainly appears to be too hard for many people when confronted with dissent from the correct opinion. The result has been obtained so often that it seems to be established beyond reasonable doubt.

There are two potential reasons why people behave in this way. First, someone may genuinely believe that the group – other people – have better information and so decide that it makes sense to copy their decisions. In the classic example of going to eat in a strange city, the rational choice is to select the restaurant which looks busy rather than its empty near neighbour. The diners may very well have information that it is indeed the better of the two. But, as we have already seen and will revisit below, there is no guarantee that this is necessarily true. The second motive is that people might well have a simple desire to conform, especially when the incorrect view is expressed firmly and consistently by the majority.

Copying the majority is certainly a strategy widely adopted in non-human worlds. Think of fish. Juicy fish such as herrings, which large predators relish. Such creatures have evolved the shoal or swarm as the natural way of moving around. One potential reason for this is the so-called confusion effect. A predator confronted by a large number of prey may experience sensory overload and so be less able to single out a single victim to gobble up.

*

Copying as a strategy can certainly have its drawbacks. Dirk Helbing of ETH Zürich is an intellectual leader of the social physics community – scholars trained as physicists who have turned their attention to such social and economic questions as how to evacuate a sports stadium safely in an emergency. Helbing and his colleagues have shown that simply providing more exits at regular intervals, a common-sense approach, may not necessarily work. And the reason for this is the principle of copying. In an emergency situation, an individual usually cannot gather much information, and so may believe that the group is better informed. But it is precisely this which leads to the phenomenon of flocking, so that large numbers may all try to use a single exit, whilst others may remain at best only sparsely used.

A completely different example, this time of successful copying, is provided by the rice farmers of Bali. The Indonesian island is far more than a resort for sun-worshipping Westerners. It sustains an exceptionally productive rice-farming industry. Essentially, Bali is a large extinct volcano with a lake in the crater that provides much of the water used to irrigate the crop.

For centuries, the rice growers of Bali have been engaged in agricultural practices which have proved highly successful. Without centralised control, the farmers have evolved a carefully coordinated system which allows productive farming in an ecosystem rife with water scarcity, pests and diseases, and apparent conflicts of interest between different groups of farmers. No one has planned this system. And farmers operating at different levels of the mountain have potentially conflicting interests about the control of water, the timing of rice planting, when fields should lie fallow and so on. Yet a highly productive, successful and sustainable system has evolved.

There are two key reasons why this has happened. First, the farmers simply copy the good behaviour they observe in others. Second, they place great emphasis on purely voluntary structures which promote and reinforce this cooperative pattern of behaviour and prevent less desirable patterns of individual behaviour from cascading across their social networks.

Top anthropologist Steve Lansing of the University of Arizona and the Santa Fe Institute has a very neat model which accounts for the key empirical features of Balinese agriculture. A crucial reason for the success of the system is that the pattern over the course of the year in terms of which fields lie fallow is very effective in controlling pests. This pattern can be explained almost entirely by a very simple model. Farmers operate in small units called subaks. Suppose we start the model off with each subak being allocated at random a particular month of the year for cropping. Bali lies almost on the equator, and its climate is very stable, enabling rice to be planted in every month of the year. At the end of the year, each subak simply observes the crops of its four nearest neighbours. If one of these has a better crop, the subak copies the timing of that neighbour for next year's cropping. The model rapidly converges on an overall pattern which is very similar to that which is actually observed. The subaks do not have masses of quantitative data, they do not perform the complicated mathematical 'optimising' procedures of economics, they do not rely on government officials and rules and regulations. They simply copy the good behaviour of their neighbours. Copying, simple imitation, works.

The second key reason for Bali's success is the strong emphasis placed on mechanisms for promoting cooperation. Within each subak, an individual farmer has incentives to free ride on any group arrangements. To share in the group's rewards without making a contribution. Pure short-term individual profit maximising suggests that individuals *should* behave in this way. But they don't. Great emphasis is placed on the social structure, on reinforcing the idea of individuals belonging to the subak. In other words, social conventions have evolved to try to restrain the human urge, the instinct, to react to incentives. Within the subak, of course, social sanctions could ultimately be taken against a non-cooperative individual. Outside the subak, this is not possible. Instead, an elaborate system of 'water temples' has evolved, which facilitate voluntary coordination and cooperation. The purpose is to make the social network robust, to prevent cascades of non-cooperative behaviour from spreading.

In the 1970s, the modern planner stepped into this world. On the advice of the Asian Development Bank, the government of Bali legally mandated the introduction of new, high-yielding varieties of rice. Armed with the modern panoply of mathematical models and plans, the government brought explicit regulation into this voluntary system. It was a disaster. The existing coordination mechanisms amongst farmers broke down and pests proliferated. Attempts to control pests by introducing yet further new varieties of rice resulted in the emergence of new pests. The whole planned, regulated experiment failed. It had the best of intentions, but the outcome proved dramatically different from these intentions. The idea was, fortunately, abandoned and agriculture on Bali reverted to the voluntary, unplanned but very successful system which had evolved over centuries.

*

In general, copying is usually a very good strategy to adopt. Kevin Laland, an evolutionary biologist at the University of St Andrews, posted a question on a website in 2007 and emailed it to academic departments around the world: 'Suppose you find yourself in an unfamiliar environment where you don't know how to get food, or travel from A to B. Would you invest time working out what to do on your own, or observe other individuals and copy them? If you copy, who would you copy? The first individual you see? The most common behaviour? Do you always copy or do you do so selectively? What would you do?'

In essence these are the questions posed by this entire book. Do we act as an economic rational agent and work things out by responding to the pay-offs yielded by different strategies, different choices? Or do we just copy other people? But not just this. Who do we copy and when? We have already seen examples where people copy others seemingly at random, and where, as in the case of the Protestant martyrs, the prestige of the individual is a key factor in whether or not they are copied.

The eventual outcome of the question posed by Laland was a paper published in the world's top scientific journal, *Science*, in April 2010, with nine co-authors, including UCLA's Rob Boyd, perhaps the world's greatest authority on cultural evolution. The article reports the results of a 'social learning' tournament – social learning being essentially the jargon in these particular academic areas for copying – in which entrants submitted strategies specifying rules of behaviour on how and when to use either social learning or individual trial-and-error learning to acquire successful behaviour in a complex environment. More than a hundred teams from around the world competed and submitted strategies in this computer-based tournament.

The problem the strategies – let us think of them as players, even though the 'player' here is lines of code specifying how to make choices – were confronted with is the so-called 'restless multi-armed bandit'. The idea is based on the one-armed bandit slot machine, in which pulling a lever sets off a game of chance with outcomes which yield different pay-offs. The pay-offs themselves change randomly over time, hence the word 'restless' in the description. The winner would be the player acquiring most points from these pay-offs over a long period of time.

Each player has three options in every round: *observe*, which simply means watching another individual and the pay-off this agent obtains; *innovate*, which involves developing a behaviour of its own; or *exploit*, which involves actually invoking one of the strategies it has acquired and pulling the arms of the bandit. Only by exploiting can players acquire points. So an important decision for players is whether to stick with the strategies already acquired and 'exploit' on every round, or whether to invest time in learning a new behaviour which may or may not be better than the ones it already has.

As noted above, the tournament was made even more realistic by the fact that the pay-off point associated with each behaviour was not fixed, but varied over the course of the tournament. Strategies which had yielded lots of points might suddenly yield many fewer. This is a key element of the real world. Companies innovate all the time to try to outperform their rivals, so the competitive environment is constantly changing. For example the printed book revolutionised the world when it first appeared over 500 years ago. The skills of those involved in transcribing the written word by hand, the previous dominant technology, soon became redundant. And now the printed book itself faces serious competition from electronic media.

The design, testing and refinement of the tournament took eighteen months. The play took over a year, and was only feasible thanks to no less than 65,000 hours of free computing time being provided by National Grid, the company which owns the British gas and electricity transmission systems. The experts involved in designing the tournament had clear views in advance as to the type of strategy which would succeed. As the abstract of the *Science* paper states: 'Most current theory predicts the emergence of mixed strategies that rely on some combination of the two types of learning.' In other words, players would use a mixture of gathering private information by trying out their own strategies and seeing what pay-off they got and using public information by copying the behaviour of other players.

To their astonishment, a very simple strategy devised by two Canadian post-graduate students, Timothy Lillicrap and Daniel Cownden, proved the outright winner. And this strategy relied almost exclusively on copying. To quote again from the heavy-duty paper: 'This outcome was not anticipated by the tournament organisers, nor by the committee of experts established to oversee the tournament, nor . . . by most of the tournament entrants.' At one level, copying may seem advantageous because agents can avoid the costs of trying to work things out for themselves by rational learning. However, it may result in the acquisition of an inappropriate or outdated strategy, and for this reason the expert view was that it would be used sparingly by the tournament winner.

A common feature of the most successful strategies as a group was the emphasis which they placed on copying. But not only that. They spent as much time as possible 'exploiting', acquiring points, rather than searching and investing time in acquiring even better strategies. So they did not attempt to acquire an 'optimal' strategy, in so far as this word has meaning in the context of a changing environment. They found something which gave them points and used it. In other words, they took advantage of positive linking. The network in this case was the other strategies whose behaviour and success they observed. They copied, and they won.

Social learning – 'copying', in our terms – appears to be a very deep-seated aspect of human behaviour. Evolutionary anthropologists and psychologists have argued persuasively that the anomalously large brain (neocortex) size in humans, compared to other mammals, evolved primarily for social-learning purposes. As Robin Dunbar and Susanne Shultz of Oxford University's Institute of Cognitive and Evolutionary Anthropology stated in a paper in *Science* in 2007: 'The evolution of unusually large brains in some groups of animals, notably primates, has long been a puzzle. Although early explanations tended to emphasise the brain's role in sensory or technical competence (foraging skills, innovations, and way-finding), the balance of evidence now clearly favors the suggestion that it was the computational demands of living in large, complex societies that selected for large brains.' So the skills required for positive linking may even be hardwired into the human brain.

Of course, true believers in economic rationality have no real difficulty in reconciling the outcomes of the tournament with their view of the world. *Any* mode of behaviour which proves the best is obviously optimal and therefore would be learned by an economically rational agent. We might even say, invoking the old Soviet phrase again, that it is apparent that this is indeed the case. However – and this point is crucial – using copying as the main way of making choices can lead to outcomes which are so radically different from those predicted by rational economic behaviour that it really does not make sense to describe the former as 'rational' in the economic sense of the word at all.

The social-learning experiment introduced yet another, even more realistic aspect of the real world. The network which any given agent used to help make decisions was not fixed in structure, it was not static. It evolved over time. The group of individuals being monitored for their performance varied over time. The dynamic nature of the network is yet another way in which network approaches are in general much better approximations of reality than the models of orthodox economics. And in such worlds – both model worlds in the computer and the real human world – our understanding of what it is to be rational changes completely.

Incentives matter in the experiment; indeed, they are a key feature. You win the game by acquiring more points than anyone else, and you acquire points by making a series of decisions over time. But there is no meaningful sense in which we can say that even the strategy of the winning algorithm, the winning player, was 'optimal'. It was the best at exploiting the information provided by the networks of other agents, but that is the limit of what we can say, even ex post, after the experiment has been carried out. And ex ante, before the event, the most successful strategy in the experiment was extremely difficult to identify, a fact attested by the failure of some of the world's leading experts involved with the design of the experiment to anticipate that copying strategies would prevail.

So, yes, incentives still feature. But whether they are the conventional incentives of price – points in the experiment equate to price in the real world – or the more sophisticated range of incentives falling under the banner of 'nudge', the economic concept of rational behaviour, of responding to situations in this way, really makes little or no sense in this evolutionary, network context. Behaviour based on copying is the rational way to behave, a mode of behaviour with dramatically different implications from those of conventional economic rationality.

*

How much difference can the copying motive make to the outcome? When an important determinant of agents' choices is simply the proportion in which other agents have already selected the various offers, what happens compared to when they act 'rationally' in an economic sense?

Duncan Watts, the Columbia professor who moved to Yahoo! and whom we have already met, and his colleagues Matt Salganik and Peter Dodds carried out an intriguing experiment on this, which they published in *Science* in 2006. They created an artificial music market comprising 14,341 real participants, recruited mostly from a teen-interest website, who were shown a list of previously unknown songs from unknown bands. On arriving to take part in the live experiment, the young people were divided into one of two groups. The groups differed only in the amount of information provided to them when they were making their selections. In each case, each individual made choices on his or her own, without any others being present. A choice in this context was defined as being the decision to download a particular track.

Given the name of the song and the band, one group were allowed to listen if they wished, and were asked to assign a rating from one star (hate it) to five stars (love it). There were forty-eight songs in total from which to choose. Once someone had listened to a song and rated it, he or she was given a chance to download it, though this was entirely optional.

This part of the experiment essentially corresponds to the world of economic rationality. Individuals are choosing quite independently. They are given complete information, the names of the songs and the bands, and can listen to the songs before deciding whether or not to download. The amount of choice is not excessive, and the products are easy to evaluate. The teenage participants were not asked to decide on the relative merits of different renditions of Brünnhilde's great final aria in Wagner's *Götterdämmerung*, but to choose from a genre of music with which they were very familiar. Of course, the tastes of the participants may very well have been shaped prior to the experiment by the preferences and choices of their peers. But within the strict confines of the experiment, we can take their tastes as being fixed.

The number of individuals involved in each experiment varied, but it was usually between 500 and 1,000. At the end of each experiment, the numbers of downloads of each song were totalled. This distribution can plausibly be regarded as the independent preferences of the members of the group. Figure 7.1 plots the relative popularity of the forty-eight songs in a typical experiment. It shows how many times each song was downloaded, the totals across the individuals who took part. The only twist in the chart is that, in order to make comparison easier with the Figure 7.2, the actual numbers of downloads are not shown. Rather, the chart shows the number of downloads compared to the average number across all forty-eight songs. For simplicity, the average is set equal to 100.

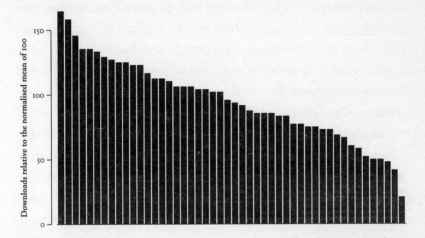

Figure 7.1 Typical outcome of the music download experiments; number of times each of forty-eight songs is downloaded over the course of an experiment, participants know only the names of the song and band and can listen to songs before deciding whether or not to download

Note: The average number of downloads is set equal to 100 for comparative purposes

The advantage of comparing everything to an average set at 100 might now be clearer. We can see that the least popular song got only about twenty-five downloads, or a quarter of the average. This stands out, however, and we can see that apart from this there are several in the 'least popular group', as it were, with around fifty downloads each, or around one half of the average. In contrast, the most popular get just over one and a half times the average number.

So there are genuine differences in popularity of the forty-eight songs, which can be reasonably seen as reflecting the individual preferences of the agents in the experiment, because each person's preference was recorded in isolation. Broadly speaking, the most popular are around three times as popular as the least.

We now move to the second group of individuals selected by Watts and his colleagues. In every respect but one, the experiment was exactly as described above. The difference was that each participant could see how many times any particular song had been downloaded already. At the start of each experiment, the clock was reset to zero, so people could only see the choices in their particular experiment, and not those of any other experiment. In some of the experiments, the songs were simply listed in alphabetical order, and the previous downloads set against the name. In others, the songs were listed in order of popularity, a hit parade with the existing number one at the top. Figure 7.2 shows the results of a typical experiment when agents have this one extra piece of information, with the songs listed in order of popularity.

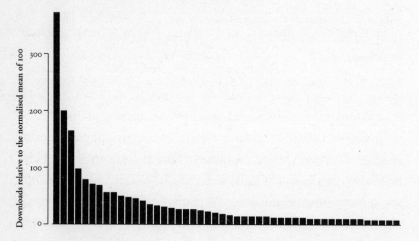

Figure 7.2 Exactly as Figure 7.1, except the participants know the number of previous downloads of each of the songs before they decide themselves

We do not need arcane statistical theory to know that the distribution of outcomes is completely different in the two figures. In the second case, most songs end up with very few downloads. The most popular gets around 350, the second most popular 200. The ratio between the most popular few and the least popular song is not just three to one. It is at least thirty to one. As Watts and his colleagues point out, the experiment by its very nature limits the degree of inequality of outcome which can emerge. The number of choices is very limited, for one thing – forty-eight to be precise. And participants could not meet or otherwise confer about the songs as they would be able to do in real life. They just had a single piece of information, no more and no less, about the choices which other people had made. But even so, compared to the outcomes under independent preferences, there was a very marked increase in the inequality of the various downloads.

But the results are even more dramatically different than this once we lift up the lid. The most popular are considerably more popular than those in Figure 7.1. But they are not the same! The most popular song in each of these two charts is different.

Giving people this one extra piece of information, on how fashionable each track is, does not make already popular ones more popular, and those which are less liked to be chosen even less. It alters spectacularly which songs end up as the most or least popular. Duncan Watts and his colleagues concluded that there was some relationship between the relative popularities of songs in the two types of experiment. But it was very weak. 'The best songs rarely did poorly, and the worst rarely did well, but any other result was possible.' So songs on the far right of Figure 7.1 did not end upon the far left of 7.2, and vice versa. But any other result was possible! A song in the middle of Figure 7.1 could end up in 7.2 either being wildly popular or, more likely, getting almost no downloads at all.

Where fashion matters, it affects the outcomes dramatically. Is it possible to use this knowledge to create self-fulfilling prophecies? Watts and Salganik went on to consider this question in *Social Psychology Quarterly* in 2008. They note an intriguing real-life example reported by American economist Alan Sorensen. He noticed that there were occasional errors in the construction of the *New York Times* best-seller list, and found that books mistakenly omitted from the list had fewer subsequent sales than a matched set of books that correctly appeared on the list.

But the main point of their paper essentially replicated their earlier experiment. With one difference. During the course of any particular experiment, at some point the relative ranking of popularity settles down, certainly as far as the most popular are concerned. A sufficient gap emerges between the number of downloads of the number one and two, two and three and so on, to make it difficult for the order to subsequently change.

At this point, Watts and Salganik gave false information to subsequent participants in the experiment. And the information really was false. They did nothing other than to completely reverse the rankings which had emerged. So the next player was told that what had actually been number 1 was number 48, number 2 was number 47, and so on. Most songs did in fact experience self-fulfilling prophecies, so that previously unpopular songs went on to become popular. The inversion did not hold true completely, for the songs which genuinely were the most popular in the independent-agent version of the experiments did gradually regain popularity if they were falsely reported as being way down the list. And participants tended to make fewer overall downloads when they were told that unpopular songs were popular. But false information on fashion still made a difference, even when it was by deliberate design the complete opposite of the truth.

*

Enormous resources are devoted to the task of predicting the out-come of social processes in domains such as economics, public policy, and popular culture. But these predictions are often woe-fully inaccurate. Two striking characteristics of popular cultural markets are, first, inequality (hit songs, books, and films are many times more popular than the average) and, second, unpredict-ability. In the original Watts experiment, songs which were well regarded by the players of the game when choosing independent-ly rarely came near the bottom when the network of the choices of others was allowed to operate, and poor songs rarely came near the top. But, and this remarkable phrase certainly bears repeat-ing, any other result was possible.

Imagine, then, that we are in the marketing department of a large corporation, developing a new variant of an existing brand. Let us say, fruit-coated Mars bars (which have never existed, as far as I know). The process would follow the same course as a political party or a pressure group testing out different nuances of a potential policy change. We experiment with a number of different alternatives – cherry, banana, gooseberry – and carry out extensive testing of the products with consumer groups. This may enable us to screen out fruits which nobody seems to like and allow us to settle on a smaller group, each of which reso-nates to some extent with potential consumers. Which of them to launch? If our market research has been mainly focused on consumers operating as individuals, the Watts experiment tells us that, if network effects matter in this market, then we cannot predict ex ante, before the launch, how well the new variant will actually perform. Most new products fail. And networks are the key reason why.

However, there is some hope. The future is not completely impenetrable. Before a product or policy or campaign is launched on the public, the curtain does remain firmly drawn. But chinks

of light soon begin to appear. An important feature of such fash-
ion-driven processes is that the choices which eventually prove to
be the most popular seem to emerge at a very early stage of the
process.

The 2010 conference on Social Computing, Behavioral
Modeling and Prediction held in Maryland attracted a bevy of
illustrious sponsors, such as the National Institutes of Health, the
Office of Naval Research and Air Force Research Laboratories.
All these outfits are aware of the inherent difficulties of prediction in a networked world. I published a paper in the proceedings
of the conference which examines the data set from the original
experiment carried out by Watts and his colleagues.* The question was: can a simple rule be devised which will enable early
detection of the song which eventually ends up as number one in
each particular experiment?

The answer is unequivocally 'yes'. And, paradoxically, the
stronger the potential effect of the random network which connects you to the previous choices of others, the more likely it is
that the winner can be identified. Recall Watts's three types of
world: independent choice; previous choices known and ordered
alphabetically; and previous choices known and ordered by popularity. Before the start of each experiment, predicting the winner, even with the knowledge of the outcomes of the independent preference experiments, is virtually impossible. We can make
some very tentative remarks about a few songs which are unlikely
to succeed and about a few which might. But that it is all.

The situation alters completely once the experiment starts.
The trick is not to examine any of the inherent qualities of any of
the songs, the way which economic rational choice would point

* The detailed description of the available data is available at http://opr.
princeton.edu/archive/

the enquirer. It is to discover which sort of world we are operating in, how powerful are the network effects. Once enough steps have been taken, enough downloads made, for us to know that we are in the 'strong' network effect version of the experiment, the song which is the current number one has a very good chance of remaining there. And the identification can be made when the number of downloads of each song is still in single figures!

An appreciation of how this can be the case is as follows. (For interested readers, the technical paper is cited in the short list of further reading at the back of the book.) In Chapter 6 we met the great mathematician Carl Friedrich Gauss, and the so-called Gaussian, or normal, statistical distribution. It is the favourite distribution of economists, with much of their statistical analysis resting on the assumption that this is how the world is. But Herb Simon noticed that in a wide range of contexts, the outcomes of economic and social processes looked completely different. In particular, they were 'skewed'. In other words, they exhibited a high degree of inequality of outcomes, with the biggest, the most popular, being many times larger than the average in ways which the Gaussian does not really permit. Such outcomes arise precisely in systems where the component parts interact with each other, where the behaviour of any given agent can be influenced directly by the behaviour of others. So if we see in practice a distribution across a set of choices which looks like this, we know we are looking at a world where networks and copying are important.

The idea that in a networked world the early stages of any evolutionary process, such as how the popularities of different choices evolves over time, are decisive was shown very neatly by Brian Arthur, a highly innovative British economist who has been based for many years in New Mexico and California. Arthur's original work on this was carried out in the 1980s, well before the explosion of network science, so it is not formulated explicitly in these terms. But it has the advantage over the kind of rule of thumb described above because he obtained analytical results for a particular type of process, one which is often found in the real world.

Arthur's initial work was on a highly abstract concept in non-linear probability theory, something called Polya urns. Several rather obscure journals have already been mentioned in this book, but perhaps the one where this work first appeared is a candidate for the most otherworldly. He wrote a paper in 1983 with two Russian mathematicians, Ermoliev and Kaniovski, in a journal with the Russian title *Kybernetica*. Imagine we have a very large urn containing an equal number of red and black balls. (The colours are immaterial.) A ball is chosen at random, and is replaced into the urn along with an additional ball of the same colour. The same process is repeatedly endlessly. Within this enormous urn, can we say anything about the eventual proportions of red and black balls which will emerge? They start off with a 50/50 split. Can we say how this split will evolve?

Indeed we can. Arthur and his colleagues showed that as the process of choice and replacement unfolds, eventually the proportion of the two different colours will always – always – approach a split of 100/0. It will never quite get there, because at the start there are balls of each colour, but the urn will get closer and closer to containing balls all the same colour. The trouble is, we simply cannot say in advance whether this will be red or black. Just as in Duncan Watts's experiments, predicting the eventual winner before the process starts is extremely difficult. Except that in this context, the formal solutions to Arthur's equations show that it is literally impossible to do any better than a pure 50/50 guess. They also show that the winner emerges at a very early stage of the whole process. Once one of the balls, by the random process of selection and replacement, gets ahead, it is very difficult to reverse. The abstract nature of the analysis does not enable us to say what exactly constitutes 'very early', as we can do with rules of thumb in empirical contexts. But it establishes the principle unequivocally.

What has all this to do with the price of fish (or anything else, come to that)? Concepts of balls being drawn and replaced at random from an infinite-sized urn do seem rather abstract. We can usefully think of the balls as being connected on a network, one which evolves over time. As soon as the first ball is replaced, the chances of drawing one of the same colour on the next draw have become ever so slightly better than 50/50. We might need a large number of decimal places to say how much better, but better it has certainly become. It is as if a consumer is choosing between two complex alternatives, and is paying no attention to the inherent qualities of the balls, their colour, but just to how popular they were. The distribution of balls when *you* come to choose reflects the random networks which connect the people who have already chosen.

Brian Arthur explained the practical significance of his results in an article for the *Economic Journal* in 1989. Suppose a new technology emerges – internet search engines, to use a post-Arthurian example – and consumers have two alternative versions of the technology from which to choose. Arthur used the example of video recorders, the Sky+ box of the 1980s. The technology is entirely new, so no one really knows how to evaluate the various products associated with it. The principle of copying seems entirely rational. Someone makes a choice. In the abstract model, this is the extraction at random of a ball from the urn. The fact that brand A has been chosen rather than brand B tilts ever so slightly the possibility that the next choice will also be A rather than B – this is the replacement rule in Arthur's model. And so the process unfolds.

In practice, of course, as one of the brands gains a lead over its rival, factors other than consumer copying will come into play and reinforce its dominance. There will be positive feedback, positive linking, so that success breeds further success. The more successful brand may be able to advertise more, for instance. Retailers will give more shelf space to it, and may even, in the splendid language of retailers, delist its rival, so that it becomes harder and harder to obtain. Technologies and offers which piggyback on the brands – think of apps and iPhones here – become increasingly designed to be compatible with the number-one brand.

An important implication of all this, again completely consistent with Watts's experiments, is that success is not necessarily related to quality. In the traditional model of rationality, provided that the information can be supplied to consumers (or voters), they can be relied upon to select consciously, to choose the alternative which is best for them. In a world where the principle of copying is rational, for whatever underlying motive, this no longer holds.

*

We are beginning to get the pieces of the jigsaw assembled for a model of rational agent behaviour which is relevant to the world of the twenty-first century. Incentives still play a role, but the most important aspect is what we describe in shorthand as 'copying' the behaviour of other agents in a network. The mathematical features of networks and how they can impact behaviour were discussed in the previous chapter. In this chapter, we have fleshed out the concept. Psychological evidence suggests that copying is a very sensible mode of behaviour to adopt in the modern world. The amount of choice is enormous, the products and services available are often complex and difficult to evaluate, and giant leaps in communications technology mean that we are much more aware than ever before about what other people are doing, thinking and buying. The concept of Rational Economic Man has become largely irrelevant. Instead, we have Rational Copying Person.

This model of behaviour, which is able to explain a wide range of features in human social and economic systems, has quite different implications from those of the economic model of rationality. And these implications are often disturbing. Unequal outcomes – skewed distributions – appear to be an inherent feature whenever network effects are important. Network effects seriously weaken the connection between the qualities, the attributes, of any set of alternative choices and the relative frequency with which people make these choices. And network effects increase the uncertainty policy makers face when addressing any given problem.

But we have also seen some pointers which can help rather than hinder policy makers. When network effects are important, even small changes brought about by changes in incentives can be magnified dramatically. Networks can make policy more rather than less effective. Positive linking has the potential to transform the world for the better.

There is a rule of thumb which is a pretty good indicator of whether network effects are present in any particular situation. The structure of the network, its type, offers good guidelines to policy makers on how to encourage or prevent the spread of particular behaviours, the cascades, across networks of agents. And there may very well be reasonable early warning signs in any situation in which network effects are important on whether a policy change is likely to succeed or fail.

These are all points to which we return in the final chapter. But first there is one outstanding point to address, one issue to examine, in this twenty-first-century approach to rational agent behaviour. As we have seen, network effects give rise to unequal outcomes. A particularly disturbing feature of such a world is the self-reinforcing nature of the processes at work. This seems not only to ensure great inequalities of outcome, but outcomes in which the number one stays there for ever. And, by implication, those choices, whether of products, ideas, lifestyles, which are doing less well at any given time appear condemned to have no chance at all of ever succeeding in the future. Are there more subtle versions of the rational copying model, versions which both enhance its empirical realism and perhaps qualify some of its very stark implications? It is to this which we now turn.

8

All Good Things Must Come to an End

Network effects take us into territory where outcomes are distinctly unequal. Herb Simon's article on skewness, as we have seen, languished relatively unread for many years. But we now see it as a most prescient paper, decades ahead of its time, describing the basic mechanism which underlies the emergence of the unequal outcomes we observe in the real world of human social and economic systems in a wide range of disparate settings.

Rational behaviour in the twenty-first century is essentially based on this approach, on 'copying' the decisions of other agents. The word 'copying' is again put in quotation marks, to remind us that it is a shorthand description of a range of plausible and realistic motivations for such behaviour. This approach enables us to account for a wide range of real-world outcomes.

But, as discussed at the end of the previous chapter, the approach implies that we get locked into these unequal outcomes. Once the differences between choices, between alternatives, starts to widen there is no mechanism for reversing the trend. The rich get richer, more popular choices become even more popular, and those left behind appear to have no prospect of future success. They seem condemned to failure. Yet this does not seem to be what actually happens. Yes, we see inequalities in outcomes, which require us to invoke network effects rather than incentives in order to be able to understand them properly. But there does seem to be turnover in the outcomes over time. The most popu-lar, the most successful, the biggest, does not stay there for ever.

Consider a major contemporary problem: unemployment. As I write these words in the summer of 2011, after the unemployment rate in America has experienced a small but nevertheless distinct fall from its peak level of 10.1 per cent in October 2009. But the rate is still high, and it is a big political issue. This figure measures the percentage of the people in the labour pool who are registered as unemployed. But the true proportion of those excluded from work is even higher. Almost all Western countries operate welfare systems in which out-of-work individuals are not classified as unemployed if they are in receipt of certain other benefits, such as disability benefit, irrespective of whether they are capable of work. Some may regard this state of affairs as merely a cynical attempt by governments of all persuasions to try to disguise the true total of unemployment, others may regard it as socially beneficial.

Regardless of what interpretation we place on this, and regardless of whether we focus on the registered unemployed or the wider population of adults of working age not in work and on some sort of benefit, we observe very wide disparities of outcomes at local levels. This is the case regardless of the overall state of the economy and the overall level of unemployment. In good years, the average number of those not in work is lower than in bad years, but we still see great inequalities in local outcomes.

A social housing scheme in the town where I was born attracted national notoriety in the UK in 2010, when it was discovered that a whopping 84 per cent of all adults of working age living on it were not in work and drawing some sort of benefit. Yet there were almost identical schemes in the town that had much lower rates. The rates were still high, both because there had just been a major recession and because a large proportion of the relevant populations were unskilled, but they were 20 or 30 per cent rather than 84. It should be said that the entrepreneurial spirit in the area is not entirely dead. A cafe on the scheme promoted on its outside board an 'All-day full English breakfast for £4.50 – with a can of Stella'. Even without the lager it wasn't bad value.

We see unequal outcomes in unemployment rates at local levels, but we also see turnover in the relative performance over time. The basic unit of local government in America is the county, of which there are just over 3,000 in total. Their sizes depend on local political factors and vary enormously, from the giant Los Angeles County to tiny rural ones with mere hundreds of people. On average, an American county is of similar population to the average local authority in the UK. But their size is not the immediate topic; rather it is the distribution of their unemployment rates. The rates tell us the number of unemployed, as a percentage, compared to the total workforce.

If we go back twenty years to 1990, the economy was pretty sluggish, heading into the mild recession of 1991. The average unemployment rate across the counties as a whole was 6.2 per cent.* But the lowest unemployment rate in an American county was just 0.5 per cent, in Grant County, Nebraska. In case this might just be a fluke because the county itself is tiny, there were no fewer than 205 counties where less than 2.5 per cent of the labour force was unemployed, the lowest rate ever recorded at the national level, way back in 1953. In contrast, the highest was an enormous 40.8 per cent, in Starr County, Texas. There were 265 counties with rates above 10 per cent, the peak reached in the most recent recession in 2009.

Flash forward now to 2010. The average rate of 9.4 per cent

---

* This is the average across the counties, regardless of their size, so it is not exactly the same as the unemployment rate across America as a whole.

across the counties was higher than in 1995 because of the impact of the financial crisis. But there was still a wide disparity of outcomes. The lowest, 1.6 per cent, was Slope County, North Dakota, where it seems like the recession never happened. In 115 counties the rate remained below 4 per cent, the forty-year national low observed at the height of the dot.com boom. The highest was 27.6 per cent, in Imperial County, California, and in 2010 the rate was above 15 per cent in no fewer than 128 counties.

There was certainly a fairly strong tendency of counties with relatively high or low unemployment rates in 1990 to also have high or low rates, compared to the average, in 2010. For those of a statistical bent, the simple rank correlation was 0.60. But, clearly, the rankings do not stay fixed. There is turnover. So, just look- ing at counties with substantial labour forces rather than the tiny rural ones where small changes in jobs can generate big changes in unemployment rates, examples more or less at random are Union County, New Jersey, with a workforce of over 250,000, which fell from rank 1,365 in 1990 to 1,775 in 2010 and Du Page, Illinois, with more than half a million workers, which moved from rank 664 to rank 1,238. Big counties which improved their relative positions include Plymouth County, Massachusetts, with over 200,000 workers, which rose from 2,200 to 1,560, and Bexar County, Texas, with more than half a million workers, which improved dramatically from rank 2,279 to 879. Even the oft- derided Baltimore City improved its relative position, moving from 2,436 to rank 2,262, still a bad rank to have but nevertheless moving in the right direction.

A clear example from the British labour market is provided by the local areas in the UK where coal mining used to be impor- tant. For many years regarded as the elite of the British industrial working class, the miners were treated with kid gloves by suc- cessive governments. Harold Macmillan, a Conservative Prime

Minister in the 1950s, once said that there were three bodies which no sensible politician would ever challenge: the Catholic Church, the Brigade of Guards and the National Union of Mineworkers.

All this changed in the early 1970s. The country had another Conservative Prime Minister, Edward Heath. The miners were becoming more and more militant, symbolised by the rapid rise of a charismatic hard-line left-wing union leader, Arthur Scargill. Unlike Macmillan, whose social roots were firmly in the upper class and the tradition of noblesse oblige, Heath had risen from the petit bourgeoisie, with its rather more Poujadist attitude towards the workers. He decided to teach the miners a lesson. The outcome was Heath's total humiliation. The miners physically overwhelmed a large force of police who were trying to keep open supplies to a major coal-fired power station. Industrial disputes escalated. Heath called a general election on the platform that it was the elected government which ran the country and not the NUM. This time, Macmillan was proved right. Heath was summarily ejected from office by the voters.

It is 1979, and Mrs Thatcher enters the scene, determined to crush the union once and for all. She waits patiently and chooses her moment carefully. Coal stocks are gradually built up during the next five years before a series of provocations are launched against the miners. They take the bait and go on strike. But this time, their enemy is far more determined and there is no hint even of compromise, let alone surrender. The strike is broken in the winter of 1984–5. A short while later a major programme of pit closures is introduced, and by the end of the decade there are hardly any mines, or miners, left.

In the year prior to the fatal strike, 1983, there were twenty-nine local authority areas in the UK, out of a total of 459, where coal mining accounted for more than 10 per cent of total employment. How did these areas react to these massive shocks on employment loss and what was the outcome? Twenty years seems a sufficient period for the various feedbacks to work their way through and out of the system. So we can compare total employment in each of these areas in 2002, twenty years on, to what it was in 1983.

Across the mining areas as a whole, the effects of the shock of the pit closures persisted. In Britain as whole, employment grew by 23 per cent, but by only 9 per cent in the former coal-producing regions. But the striking feature is the sheer diversity of experience in the seemingly devastated areas. In no fewer than ten out of the twenty-nine, employment in 2002 continued to be lower than twenty years previously. In three, Wansbeck and Easington in the north-east of England and Cumnock and Doon valley in the west of Scotland, the fall was nearly 20 per cent. Yet in others, employment grew faster than the national average, with one, South Staffordshire, posting a rise of no less than 45 per cent.

We have here what we might usefully think of as dynamic tension. Tension between the self-reinforcing factors which keep a local area trapped with high unemployment, and those which might help it escape and move up the rankings. Imagine an area where there is a substantial employer. We do not have to go so far as to imagine the classic 'company town', such as Consett in the north of England, once dominated by its steel works, or Winston-Salem in North Carolina, where much of R. J. Reynolds's output was located. It is enough to have a reasonably sized employer close down, for whatever reason.

There is clearly an immediate knock-on effect on other businesses located in the immediate area as the now-unemployed workers have less money to spend. But there are more subtle signals which can reinforce even more strongly the negative feedbacks on the area. The culture of not working may spread locally, and people on low wages may gradually manoeuvre themselves into a situation where they have lots of leisure. In other words, out of work with not much income, but plenty of free time. The more people there are in this situation, the more acceptable it becomes to their immediate peers on their social networks. Social values evolve, so that being out of work may even become the new social norm. Some of the more dynamic individuals may move out and seek prosperity elsewhere. Indeed, entire firms may relocate. And once an area gets a reputation for high unemployment and lack of initiative, its image spreads across networks of companies, and it becomes less attractive for them to either expand within or move into the area.

The network approaches we have seen in previous chapters help to explain this process of self-reinforcing lock-in. They are abstract models, designed to give a general insight into these processes, and lack the richness of detail of any specific example. But both Brian Arthur's model and Herb Simon's preferential attachment describe how network effects help us understand the deep reasons why some local areas remain relatively poor over long periods of time. Through the spread of both perceptions and behaviour on the relevant networks, feedbacks are set in train which make it difficult for the area, once set onto the path of high unemployment, to escape.

Yet it is not at all inevitable that areas of relative poverty remain in this position for ever. There is turnover in the rankings. Yes, there is a strong tendency for the relative rankings of performance, on whatever context, to become fixed. But, somehow, they are not fixed. There is persistent turnover.

We see exactly the same phenomenon across almost all the wide range of examples we have encountered so far. The most popular video on YouTube today is rarely the most popular tomorrow. The song which is at number 1 this week does not usually stay very long in this position. Over the entire period from 1952 to

2006, no fewer than 29,056 songs appeared in the Top 100 chart in the UK. Of these, 5,141 were in the chart for just a single week. Almost exactly a half stayed in for less than a month, so four weeks was the typical life span, as it were, of a song in the Top 100. In contrast, fifty-nine remained popular for more than six months, and one, 'My Way' by Frank Sinatra, spent an incredible 122 weeks in the chart.* The typical life span at number 1 was just two weeks, the longest unbroken reign being sixteen weeks ('(Everything I Do) I Do It For You', by Bryan Adams in 1991). In popular cultural markets such as these, turnover is rapid.

At the other extreme is the ranking of the world's largest cities. Mike Batty, a distinguished spatial geographer at University College London, published an analysis of this in *Nature* in 2005. He begins his work with the largest US cities from 1790 to 2000. At any point in time, a snapshot of the relative sizes of these urban areas reveals a considerable degree of stability. The absolute sizes of the populations grew enormously over time as the US itself expanded, but the skewed nature of the plot stayed remarkably similar.

Such apparent stability disguises substantial turnover. Over the 210-year period, 266 cities were at some stage in the top hundred. From 1840, when the number of cities first reached one hundred, only twenty-one remain in the top hundred of 2000. On average, it takes 105 years for 50 per cent of cities to appear or disappear from the top hundred, whilst the average change in rank order for a typical city in each ten-year period is seven ranks. This latter point means that if a city is, say, number 50 in size now, on average in ten years' time it will either be number 43, seven places higher, or number 57. It may of course be neither; this is the typical experience.

In terms of world cities, using the data from 430 BC mentioned

* I absolutely prefer the Sid Vicious version, it has more zest.

a couple of chapters ago, the turnover in the top fifty is slower, but it is still there. Naturally, the period since the Industrial Revolution, almost a blink of an eye in the overall context, is a period of rapid transition, but other periods saw distinct changes. The pace of change has itself been variable, but it has always been there. Who would guess the longest period a city has spent in the world's top fifty or, even harder, the name of the city itself? Suzhou in China records 2,158 years in the biggest fifty cities of the world, closely followed by Nanking with 2,080 years. No city in the top fifty in 430 BC survives in the list in 2000, but of the fifty biggest at the time of the Fall of Constantinople in 1453, a major event in world history, six are still there.

Team sports is another area where unequal outcomes combined with persistent turnover is observed. In American football, the champions of the two major conferences, the NFL and the AFL, play each other for the Super Bowl. These are the elite teams of all those who play the sport. In the forty-five seasons of its existence, the Super Bowl has been won by no fewer than eighteen teams. But just four of them – Pittsburgh Steelers, Dallas Cowboys, San Francisco 49ers and Green Bay Packers – account for twenty of the victories, so there are unequal outcomes. And even though a team like the Steelers in the mid-seventies or the Cowboys in the early nineties may appear near-invincible, eventually they drop down the rankings.

American football at its highest levels has rules designed to try to promote turnover in success. The weaker teams in any particular season get priority in signing the star college players who want to turn professional. There is a strict salary cap per team, based on the revenues of the league as a whole rather than on those of the individual teams. This further restricts the ability of the most successful sides to consolidate their success by paying their players more money than is available at the less successful outfits. Even so, there is rather marked inequality of outcome in terms of the teams which have won the Super Bowl.

Salary caps, which help promote turnover in success, are prevalent in the major American professional sports leagues. The National Basketball Association, the National Hockey League, and even Major League Soccer all have them. Baseball has its own variant, the so-called 'luxury tax'.

Somewhat paradoxically, Europe, often viewed as a hotbed of socialism from across the Atlantic, places virtually no restrictions on the ability of successful soccer teams to reinforce their success by spending more on players.* The European Champions League Final of 2011 was played between Barcelona and Manchester United, not only two of the most famous clubs in world soccer but two of the wealthiest. A pan-European club competition was first contested in 1956. There are several thousand professional soccer clubs across Europe, any of whom could in theory be crowned champions of Europe. Yet in fifty-six seasons, only twenty-one teams have ever held that title, less than 1 per cent of the total number of professional teams on the continent. And by now you will not be at all surprised to learn that there is a heavy concentration even within that small group of victors, with just six clubs winning a total of thirty-three times between them, the other twenty-three championships being distributed amongst fifteen other teams.

For long periods, teams appear unbeatable. The European competition was won by Real Madrid in each of the first five seasons, but eventually they were dethroned. The current Barcelona side are being eulogised as the greatest ever, but they, too, will eventually lose their number-one spot.**

---

* Restrictions have been discussed on and off for years, and it appears that some may soon be introduced. But these relate to the financial viability of clubs rather than the absolute amounts they spend. So teams with a lot of money who are viable can continue to spend.
** Indeed, at the final proof stage of the book, they were in fact knocked out of the 2012 competition in the semi-finals.

Soccer, and in particular what has today evolved into the European Champions League, can illustrate some important themes in the 'copying' process, the process in which success breeds further success. Although the particular points are specific to soccer, the principles involved have far more general applicability. Most situations in the human social and economic worlds will have their own nuances, their own details, which are required for a more complete explanation of the outcome. But the basic driving force is the principle of copying, the principle of preferential attachment.

We could illustrate these ideas instead with the far more momentous question of why the continent of Europe, and more specifically a small number of countries in Western Europe, came to dominate the globe in a way wholly without precedent in human history. With the benefit of hindsight, we can advance several plausible, powerful reasons why this happened. But in, say, the middle of the fourteenth century, how could this have possibly have been foreseen? Ravaged by the Black Death, suffering military defeats at the hands of the Muslim Ottoman Empire, technologically inferior to China, who could have said with any confidence that this relatively small land area would achieve world domination? But this is a massive topic in its own right. The soccer story we are about to read is much more containable, much easier to comprehend, but the principles it illustrates are very general.

Yes, success is self-reinforcing. But it is often very hard to predict in advance, and so by implication it is very hard to predict exactly when a very successful agent, be it a team, company, idea, individual city, website or whatever, will cease to be successful. The example I have chosen to use is the soccer team of the town where I was born, Rochdale. Why have they never been champions of Europe? In fact, their record as a professional team has been the complete antithesis of such success, having spent all but eight seasons of their entire existence in the lowest division of the English professional game and never having won even the humblest domestic trophy, let alone the most prestigious one in Europe.

It is easy to point now and explain why. Their total annual income, for example, is not much more than £2 million, a sum which would scarcely pay the wages of a single player even in one of the more run-of-the-mill teams in the highest division, the Premier League. So they cannot attract very good players. This may seem not only trivially obvious, but of no interest to anyone without personal links to the town. It is the dramatic contrast with a nearby rival which is of interest. For barely a dozen miles away from the comparatively humble, windswept Spotland stadium where Rochdale play sits the 'Theatre of Dreams': Old Trafford, the home of Manchester United, the most successful side in the history of the English game with a turnover well over a hundred times that of Rochdale, and possessing worldwide brand recognition.

It's easy to see now why they have finished in the top three for each of the past twenty seasons, on sixteen occasions in the top two, but both teams had equally humble origins. Manchester United started life as essentially the works team of the Lancashire and Yorkshire Railway company, based in – and called – Newton Heath, then as now a poor district in the eastern part of the city. Rochdale AFC was formed somewhat later but this team, too, began life in a heavily industrial town which was also part of the Manchester urban conurbation. And just five years prior to the emergence of Rochdale, Newton Heath had been served with a winding-up order. A consortium of local businessmen paid what in today's money is around £750,000 to rescue the club, and changed its name to Manchester United. Despite some fleeting success, the club languished, spending a good number of years in the division only just above Rochdale. In 1931 they were effectively bankrupt again and were rescued even more cheaply than before: for some £400,000 in today's money. Yet in 2011, the value of Manchester United is of the order of £1 billion!

So, in 1931, the clubs were not dissimilar. Both were based in undiluted working-class areas in the north of England, in fact in the same conurbation. One was placed a division above the other, but it was about to fold in bankruptcy. United's great rivals, Manchester City, appeared more likely to be destined for success, and indeed they won both the League and the Cup in the 1930s. Only two random events – twice being able to secure the help of local businessmen to rescue the team financially – had prevented the disappearance of Manchester United, perhaps for ever. It was not predestined that this should happen. The club did not have special qualities which led these individuals to make their investments in those desperate days during the Great Depression, but they did.

Having said this, particular factors do seem to have played a role in United's subsequent success. Just before the end of the Second World War, the club offered the position of manager to Matt Busby, a man who built not one, not two, but three extremely successful teams during the course of his career. His third squad became the first English side to win the European Champions Club trophy. But many soccer judges regard his greatest team to be the very young one which was effectively destroyed in an air crash in Munich in 1958, returning from a successful game which had progressed them to the next stage of the competition.

Once a team starts to be successful in this way, factors come into play which reinforce, though do not guarantee, its further success. It attracts more fans, more sponsors, in short more money, so it can buy better players, give them better treatment, better physiotherapy, and so on. But Busby's appointment itself was to a considerable degree one of chance. Thirty miles to the west of Manchester lies its great rival, the city of Liverpool. The antipathy between Manchester United and Liverpool FC, the second most successful English team ever, is intense. No player has been transferred directly between the two since 1964. Yet Busby almost joined Liverpool, who had been courting him for some time. The clincher appears to have been that Busby was friendly with a member of the United board through their membership of the Manchester Catholic Sportsman's Club.

Rochdale, too, following the mass migration from Ireland in the nineteenth century, has a large Catholic population. Suppose that in 1944 one of them had, in the local jargon, 'made a bit of money' and had known Sir Matt through Catholic social circles . . . Ah well, Rochdalians, whether at home or in exile, can only dream of what might have been.

The points in this story are far more general than the actual examples of the two teams. It is easy now to say why United are so successful and Rochdale less so. But, from the perspective of 1931, who could ever have predicted the stupendous success which one of these clubs would eventually enjoy? At the time, it had only just been prevented from disappearing from the face of the earth.

*

The rational model of 'copying' which we have illuminates the process by which success emerges and evolves. But it does not yet encompass why things eventually fail, or at least cease to be as successful as they previously were. Mike Batty reflects on this point from the elevated pages of *Nature* when discussing his results on the turnover in city sizes over the whole of human history. He draws an important conclusion from his analysis, couched in cryptic scientific jargon: 'The conventional model . . . cannot replicate these micro-dynamics, suggesting that such models and explanations are considerably less general than has hitherto been assumed.' What does this mean? Batty's 'conventional model' is not the conventional economic model of rational choice, but the preferential attachment model of Herb Simon, the model of twenty-first-century rational behaviour we have described. His 'micro-dynamics' refer to the persistent turnover in the rankings of cities by size. Far from converging on a stable outcome, in which the biggest just keeps getting bigger and bigger and, importantly, stays the biggest, in practice this is not true. Yes, we see a non-Gaussian outcome in terms of the size distribution at any point in time. But when we delve into the details, the micro-dynamics, we observe lots and lots of changes in an overall outcome which appears stable.

And wherever we see such non-Gaussian outcomes, we know that networks matter. We know that the potential for positive linking exists. We know that agents are taking decisions which are based, at least in part, directly on what they see or know about what other agents are doing, thinking, deciding. So unless we grasp this fundamental point, and unless we try and get some idea of what type of network it is and how it is influencing events, we will simply not have a proper appreciation of what is going on. Even a crude network-based approach is likely to give us more insight than a highly sophisticated one from which network effects are absent.

The Polya urn-based model of Brian Arthur, the preferential attachment of Herb Simon and the binary choice with externalities of Nobel Laureate Thomas Schelling and Duncan Watts – each of these models is highly illuminating when applied in the right situations. They are all based on the principle of copying rather than conscious, rational selection in the economic sense. And, as we have seen, copying is the rational way to behave in the twenty-first century. However, the outcomes of the dynamic processes at work in these models are, paradoxically, eventually static. Yes, they lead to distributions of outcomes which look like those of the real world at any point in time, but they cannot explain why there is persistent turnover *within* these outcomes over time.

Attempts have been made to modify these approaches, especially the one of preferential attachment which was rediscovered in the late 1990s by the leading physicist Albert-László Barabási, to try to take account of the universal phenomenon of turnover in rankings. So, suppose we assume that the probability of the next person choosing a particular location, brand, idea, etc., declines with the age of the object being chosen, whatever it might be. This is sufficient to introduce turnover in relative popularity of choices over time. In some contexts, this may seem reasonable, but it is a rather ad hoc extension of the basic model. Besides, ideas do not become extinct because they are old. They become extinct because no one uses them any more. Christianity is two millennia old, but it continues to flourish everywhere outside of Western Europe. Confucianism and Taoism are even older. And in many cases, the opposite is true. Some of the most frequently used words in the English lexicon are also some of the oldest.

How can we expand the behavioural model based on copying rather than rational selection to explain not only the ubiquitous right-skewed outcomes we observe, but also the persistent turnover within the rankings? We will certainly not find inspiration within economics. As Vernon Smith, economics Nobel Laureate in 2002, said in his Prize lecture: 'I urge students to read narrowly within economics, but widely in science. Within economics there is essentially only one model to be adapted to every application: optimisation subject to constraints . . . The economic literature is not the best place to find new inspiration beyond these traditional technical methods of modelling.'

I am not arguing that such a model, if it exists, would offer a complete explanation of human social and economic systems. Far from it. Not least because in many situations, as we have seen, conventional incentives play a role along with what we describe in shorthand as the copying motive. Rather, it is to establish the basis for a fundamental model of rational agent behaviour which is relevant to the twenty-first century. The traditional rational agent model of economics, with its assumptions of full information and always making the best possible choice, has partly been made more relevant by introducing limited information. In the same way, we need a behavioural model which is the new benchmark, the new basic tool we use to see how far any particular situation can be explained by it. If necessary, its assumptions can be relaxed, other factors such as incentives added to it, but it would be our first port of call in any situation which bears the hallmarks of network effects being present.

Smith enjoins us to seek inspiration not within economics itself, but from the much wider body of science. A fertile field from which to draw ideas is the work of the ecologist Stephen Hubbell, based at the University of California at Los Angeles and recipient of many honours. The biodiversity of natural systems, the focus of Hubbell's work, may seem remote from the study of social and economic issues. But there is a very good reason for asking what this can tell us. The biodiversity of natural systems has several key empirical features which should by now be familiar from the numerous examples given from a wide variety of social and economic contexts. First, a small number of species account for a substantial fraction of the total number of individuals, whatever they might be, in the system as a whole. Second, most species are represented by very few individuals. In other words, we observe the sort of skewed, non-Gaussian outcomes in biological systems that we see in profusion in the human social and economic worlds. In any given context, a few species account for most of the individuals, and most species are represented by only a few.

We can add the third key feature of such biological systems, that the rankings of the frequencies with which different species are observed changes over time. The pace of change is often very slow, because the environment in which biological evolution takes place alters much more slowly than does the social and economic environment in which humans operate. But the same principle is observed, namely that the dominant species at any point in time does not remain so for ever.

These are precisely the phenomena which we observe in the wide variety of human social and economic systems we have already seen.

*

In its modern guise, economic theory is not just remote from evolutionary, biological systems, but operates as if it were on a different planet altogether. But this was not always the case. Alfred Marshall held the principal Chair in Economics in Cambridge in the decades around 1900. As a student, he was a formidable mathematician, being placed second overall in the entire university in the final-year examinations. At his peak, Marshall dominated the discipline of economics, not just within imperial Britain, but across the world. His *Principles of Economics* became and remained the major textbook for students for much of the first half of the twentieth century.

Marshall played a central role in formalising the basic models of supply and demand within economics. But he was far from being an ivory-tower theorist. Marshall was an acute observer of contemporary economic and business life. His *Principles* are littered with insights, some elaborated at length, others merely mentioned in passing, which remain interesting and thought provoking even today. He held a persistent belief that 'the Mecca of the economist lies in economic biology'. In other words, that the economy is essentially based on evolutionary principles, rather than in the much more mechanistic concepts of standard theory, concepts which have remained in place to this day. The tools for formalising evolutionary models did not really exist in Marshall's day, nor would they for some considerable time. But he left little doubt that this is where he saw the eventual future of economics, a vision which was eliminated from mainstream thinking by the middle of the twentieth century. So, in seeking inspiration from evolution and biology, we are in distinguished company.

Hubbell developed the so-called 'neutral theory' to explain these phenomena. The theory is not without controversy but it is very influential in its field. All scientific disciplines have their own jargon, and the word 'neutral' will be decoded shortly. But, first, how does the theory work?

The basic principle on which the theory is built will by now be familiar. Imagine one of Brian Arthur's urns, but this time filled with lots of different coloured balls, where each ball is a species. The process is not quite the same as Arthur's model, not just because there are many rather than just two colours. Take one out at random, and put it back in, exactly as in the Arthur urn. But we do not add another individual of the same colour. Instead, we change an existing individual at random to be the same colour. If a species is drawn, its number increases and that of another species falls,* which makes it

---

* Unless, of course, the individual drawn at random to be changed is already the same species.

slightly more likely to be drawn again the next period.

There is an even more dramatic twist, which gives the model the ability to generate turnover. The preferential attachment rule is the fundamental one by which the system evolves. But it does not apply every single time a draw is made. Each time, there is a small probability that a different rule will be used. Namely, the individual which is drawn at random is replaced by one of an entirely new species. Essentially, both the distribution of the numbers of individuals in each species *and* how the relative frequencies change over time are described by this, the neutral theory. To repeat, because it is important, the description of the process: there is a fixed (large) number of balls of a wide variety of colours. One is drawn at random. With a given (small) probability, it is put back in, not in its existing colour, but in a colour which is not represented at all in the balls in the urn, a new colour entirely. The rest of the time, it is put back in and another ball in the urn is changed at random to have the same colour. (Alternatively but equivalently, we can think of both the ball itself and another one of the same colour being put back in, and at the same time another ball being taken out for ever.)

So, what does the word 'neutral' in Hubbell's theory mean? First consider two other words: 'rare' and 'abundant'. A plausible hypothesis is that rare species are rare because, for whatever reason, they have not adapted well to their environment. Similarly, abundant species must have particular attributes which enable them to flourish. But the word 'neutral' in this context means that *no* species has any special qualities or characteristics which make it more or less suitable to operate in its given environment. Their relative success or failure is 'neutral' to their attributes. In other words, how a species behaves, what it can and cannot do, is irrelevant to whether or not at any point in time its numbers are small or large. The outcomes which we observe are the result of purely random processes. First, the random draw of any given individual. Second, the random draw as to whether to replace another agent with one of the same kind to the one which is drawn, or whether to replace it with an entirely new species.

Of course, and this should be by now another familiar mantra, all scientific theories are approximations to reality. The neutral theory is not claimed to be accurate all the time, to explain everything. Nor is it claimed that its assumptions are necessarily completely true. But its assumptions are often sufficiently good explanations as to justify its status as a genuine scientific theory. It helps us understand the world. In any specific situation, we can always add bits of information to calibrate the model more closely to the key features we are trying to understand. But the core model, without any tweaks or add-ons, applies across a very wide range of situations.

Despite the simplicity, indeed frugality, of its assumptions, this model of twenty-first-century rational behaviour does give a good explanation of a very wide range of the social and economic outcomes we observe in the real world, from the size of the world's largest cities to downloads of popular culture such as videos on YouTube, to give two examples which both exhibit non-Gaussian distributions at any point in time, but which exhibit persistent turnover of rankings, albeit on completely different time scales.

The idea that the qualities of a species are irrelevant to its success or failure appears to strike against common sense. We might note, however, that common sense is not always a good guide to how the world actually is. It seems common sense that the earth stands still – we don't feel it moving – and the sun moves across the sky round the earth. But we know that both these common-sense perceptions are profoundly wrong.

After the process has unfolded, exactly as with the rational choice theory of economics, when we observe abundant and rare species, it is possible to tell stories about why there are lots of one and hardly any of the other. So we can describe why Manchester United is now the most famous soccer club in the world, and why Rochdale remains virtually unknown. We can give an account of why unemployment is high in one area and low in another, of why one city is successful and grows to a large size, whilst another does not. It gets much harder in many other of the examples we have seen, such as the ironing videos on YouTube. One man climbs Ben Nevis and irons. The other irons on a temporarily closed section of a motorway. One attracts more than 4,000 times as many downloads as the other, and this intrepid ironer becomes, albeit briefly, known around the world. Both at a very similar time were introduced into the YouTube ecosystem as innovations, as new species. One was much more successful than the other, for reasons which it is hard to account for in terms of rational selection.

*

Even in the much more serious context of religious faith, as we saw in the opening chapter, the eventual triumph of Protestantism over Catholicism in England in the mid-sixteenth century owed a great deal to mere chance and contingency. Queen Mary gambled that a policy of terror, of burning those who ultimately refused to reconvert to Catholicism, would succeed. The experience of the fifteenth-century Lollard heresy in England suggested it would work, and it almost did. The hard-line Protestant leaders gambled that by behaving with stoicism in the face of a terrible death, the population would be inspired by their examples, their faith would reap the benefits of positive linking. But, as we have seen, there was a strong element of chance and contingency, of randomness, in the eventual outcome. The outcomes of processes which involve copying across networks are intrinsically difficult to predict.

It is hard to argue that the Protestant success was due to the inherently superior qualities of the faith compared with Catholicism. Indeed, there were other areas of Europe in the second half of the sixteenth century where the Counter Reformation did prove to be successful, and allegiance to the Pope restored. It can, in fact, be argued plausibly that the qualities of the version of Protestantism which prevailed in England at the end of the 1550s were *less* attractive than the alternatives on offer. For this particular brand succeeded not just in undermining Catholicism, but in driving out of existence other varieties of Protestantism which were on offer at the same time. Following Luther's first defiant stand in 1517, a dazzling array of different strands of Protestantism burst out across Europe over the next few decades. Some were truly exotic, such as the Anabaptist sect under John of Leiden which established a polygamous theocracy in the German city of Munster for eighteen months in 1534–5. Communal living was a key element of the teaching, as was the use of the death penalty for almost every conceivable offence, John himself personally beheading at least one of his sixteen wives.

Most such variants sprung up, only to wither and disappear just as rapidly. But a genuine rival in England to the views of the established Protestant Church were the so-called Freewillers. The religion of the English martyrs, Cranmer, Ridley, Latimer and the rest was both sparse and terrifying. The first Prayer Book published under the young King Edward VI in 1549, for example, still contains a considerable liturgy in the funeral service, the Order for the Burial of the Dead. But by the 1552 edition, this has largely disappeared and the content is perfunctory in the extreme. Very few words of comfort are offered to the grieving relatives and friends of the deceased. This is because the then leaders of the Church of England believed firmly in the doctrine of predestination. God decided at the beginning of the world who would be saved and who was to be damned for all eternity. We might even regard this as an early articulation of the neutral theory of biodiversity! The behaviour, the qualities, exhibited by an individual in his or her lifetime are of no consequence in terms of the soul's salvation. All that matters is whether God allocated you at the beginning of time into the company of the elect. Naturally, this was a very disturbing doctrine to its followers, causing much anxiety in trying to work out into which category they had been allocated.

The Freewillers rejected the authority of the Pope and denied the corporeal presence at the Communion service. The bread and wine were, quite literally, just that; Christ's body and blood were not present. In short, they were good Protestants. But they also believed that what individuals did during their lives affected what would happen to them after death. They believed in free will, in the capacity of individuals to decide how to act. We have to be careful in assigning to people living in the sixteenth century the views of the early twenty-first century, when free will is much more closely aligned with our individualistic view of the world. But there is ample contemporary evidence which shows the anguish created by the doctrine of predestination compared with Protestant alternatives.

The Freewillers were of sufficient concern to the established Protestant leaders that, even in their prison cells under Mary, they wrote to each other about the threat posed by what a Leninist in the modern era would call an ideological deviation. John Bradford, shortly before his execution, wrote that 'more hurt will come by them, than ever came by the papists, inasmuch as their life commendeth them to the world more than papists'. In other words, the predestinarians knew that their doctrine was not attractive and that free will 'commendeth' itself much more to ordinary people.

But despite its apparently more attractive qualities, Freewill Protestantism was not the brand which was selected. The inspiration of the martyrs led to conversions not just from Catholicism but from this Protestant rival. Network effects, positive linking, offset and dominated the attributes of the two offers, and by 1560 the Freewill variant had essentially been driven into extinction.

*

So, we have a model of behaviour which assumes that people make decisions in the following way. They look at the choices which others have made and copy them, in proportion to the relative popularities of the various choices. This is the basic principle. But, in addition, they may also, with a small probability, make a choice at random. In particular, they may make an entirely new choice which no one has selected before.

To conventional thinking, it seems strange. Strange, because it makes no reference to the qualities of the choices which are available. Decisions are made without conscious consideration of the attributes of the alternatives. Instead, a deliberate decision is made to use copying. Agents are not acting in some moronic fashion, dumbly imitating the actions of choices of others. They appreciate that in complex situations, copying is the rational way to behave. It is the complete antithesis of the building block of orthodox economics, the so-called rational agent model.

But it works empirically. Most outcomes of most social and economic processes in the twenty-first century are unequal, a few alternatives are chosen many times, most are chosen infrequently. And within these outcomes, over time, relative popularities are not constant but evolve. This basic model of human behaviour is able to offer a good account of what actually happens in the real world today.

One area where this behaviour seems very applicable is in the choice of baby names. This topic may seem rather trivial, but it is a very important aspect of the culture of a society. For linguist Steven Pinker, the choice of a name 'connects us to society in a way that encapsulates the great contradiction in human social life: between the desire to fit in and the desire to be unique'. First names reveal much about a culture, including kinship patterns, popular culture trends and social values. They are found throughout time and space and are easily measured and counted. Indeed, the choices of first names reflect three principles of collective behaviour that apply to popular culture much more generally. They involve a number of people carrying out the same or similar actions at a point in time. The behaviour exhibited is transient or continually changing. And there is some kind of dependency amongst the actions; individuals are not acting independently.

The United States Social Administration provides a database on baby names that has been extensively studied in a variety of ways because of its exceptionally deep and chronologically resolved records. The data includes the top hundred baby names by US state since 1960, and, for the US as a whole, all names with at least five occurrences in each year since 1879. There have been some fascinating developments. In 1960 the most popular name for girls was Mary in most US states, except for Susan in the north-west and north-east, and some variety in the western states. This homogeneity in 1960 is also reflected in boys' names, when the five that were locally most popular (David, James, Michael, John, Robert) comprised the top five for most states, and none was lower than eighth place in any state. By 2009, this had all changed. No fewer than thirteen names were the most popular in at least one state, only one of which (Michael) had been number one anywhere in 1960. A boy's name such as Logan – the most popular name in Minnesota, Idaho and New Hampshire – was not amongst the top thirty in New Jersey or California. The same turnover and rise in heterogeneity is observed in girls' names as well. None of the 1960 names survived as 'winners' by 2009, when some of the names most popular in at least one state were not even in the top hundred in 1960, such as Madison in South Carolina and Ava in Iowa.

Anthropologists such as Alex Bentley at Bristol University in the UK and Stephen Shennan at University College London have shown that the basic 'neutral' evolutionary model of choice gives a very plausible account of how the relative popularity of baby names evolves over time. For good measure, the same article also uses it to explain archaeological pottery and applications for technology patents. With the same two authors, I have also shown, in the same journal, how the same model can explain three fundamental features of the evolution of languages. One mark of a good scientific theory is that it should be able to explain a wide variety of phenomena. Well, baby names, ancient pottery, modern technology patents and linguistic laws seem to get this one into the starting blocks!

But there are simple ways of making the approach even more powerful. In this evolutionary theory of choice and behaviour, agents look at the relative popularities of the choices already made by other agents. There is a dog here which has not barked. How far back do people look when they are making decisions? A teenager wondering what music track or YouTube video to download is not usually interested in what was popular ten years ago. Indeed, not even ten months and quite possibly not even ten days. What matters is what is trending right here and now. In contrast, a firm considering locations in which to start or develop its business often needs to take into account choices made over many years. Typically, this is the time scale on which the distinguishing characteristics of a city emerge and evolve.

The technical details need not concern us, but two quite different versions of 'memory', of how far back into the past previous decisions matter, have been developed. One assumes that only the immediately preceding period is relevant. This might be likened to genetics, in which either an existing gene is copied or a mutation happens. The system does not look back beyond the range of genes on offer right now to be copied. The other variant postulates, in complete contrast, that all previous time steps, all previous periods, are relevant. So if a variant, in whatever context we are operating, was selected even just once in the dim and distant past, it forms part of the choice set available in the present. The chances of it being chosen again are very small, but in principle this remains a possibility.

More realistically, we can allow the time scale over which previous choices have been made to vary according to the particular context. For the YouTube teenager, a short trip into the past is enough, maybe even just today's choices, and for the firm looking to expand, a much longer period. But we could select the time frame and make it relevant to any given situation rather than being required to say 'only the absolutely immediate past' or 'all previous periods' are relevant.

With anthropologist Alex Bentley and spatial geographer Mike Batty I explored the consequences of what happens when we take this realistic step.* It makes the copying approach, the model of twenty-first-century rational behaviour, even more powerful. *Any* of the skewed outcomes we observe in human social and economic systems can be captured by suitable tweaks of the two controls of the model: how often people innovate and whether they make choices other than by the principle of preferential attachment. And, when using the latter principle in order to choose, how far back do they look?

---

* The technical details are in *Behavioural Ecology and Sociobiology*, vol. 65 (2011).

The outcomes of many human social and economic systems reveal the characteristic footprint which networks invariably leave. Whenever these effects are present, whenever people make decisions, adopt behaviours, take up ideas, at least in part, by copying what others have, we know that there is a network lurking. 'Copying' is of course a shorthand description of a whole range of plausible motivations. The twenty-first-century model of behaviour explains many very diverse outcomes in the human social and economic worlds. And in our complex modern world, it is in general the rational way to behave.

We know that the effects of policies designed to influence and control such systems become much more uncertain, much more difficult to predict in advance, than in a world where most of the time people operate independently, making choices on the nineteenth-century principle of economically rational choice. In a world where there is a cornucopia of choice, of products, of lifestyles, of ideas, where many of these are complex and hard to evaluate, and where we are all increasingly aware of what other people are doing – in such a world, rational behaviour involves a strong element of copying. The outcomes and consequences of this twenty-first-century rationality are quite different.

We are at least in a better position than Aeneas in the opening book of Virgil's *Aeneid*. There, the Trojan hero encounters his own mother, Venus, disguised as a Carthaginian huntress. 'O quam te memoram, virgo?' he cries, quite unable to recognise her. 'By what name shall I call thee?' In contrast, we know the tell-tale marks of any system in which network effects operate; we know them for what they are. But does this make us any better able to control the outcomes than Aeneas himself could, to take advantage of the potential benefits of positive linking?

9

What Can Be Done

The conduct of business and policy decisions of all kinds must take account of the fact that the fundamental features of our social and economic worlds have been qualitatively transformed over the course of the past century. The word 'must' is used here with its full imperative force. Network effects are the driving force of behaviour.

The transformation has been particularly rapid during the second half of the twentieth century and in the opening decade of our current one. There is now a stupendous proliferation of goods and services available to consumers with, as we have seen, over 10 *billion* varieties on offer in New York City alone. Many of these are complex and sophisticated, difficult to evaluate even when plentiful information about them is provided. Choice is available on a hitherto undreamt-of scale, and selection amongst the choices is a challenging task.

Fashion and fads are phenomena in which something becomes even more popular simply because it has already become popular. Such patterns have existed since time immemorial. We have seen evidence of ceramic bowls being the subject of fashion in the Hittite empire, three and a half millennia ago. Success breeds success. Agents copy the choices made by other people.

The second half of the twentieth century saw a huge rise in globalisation, with the world becoming more and more open. Thirty years ago both the entire Soviet bloc of countries and

China were more or less closed to Westerners. Today, whilst Russia itself remains relatively isolated, many of its former satellites are fully integrated into the European Union, and the huge population of China is increasingly connected with the rest of the world. The same period experienced a dramatic rise in urban living. It is estimated that in 1800 a mere 3 per cent or so of the world's population lived in cities. By 1950 this had risen to around 30 per cent. It is now believed that more than half of all humanity lives in cities. In urban environments, we are far more exposed to a wide range of opinions, behaviours and choices than we are when confined to life in the village, the lot of most of humanity for most of our existence.

Developments in communications technology, based around the internet, reinforce this tendency very strongly. The internet is revolutionising communications in an equivalent way to the impact of the printing press over 500 years ago.

All scientific theories are approximations to reality. The usefulness of a theory depends upon how well its assumptions, the simplifications which any theory has to make, approximate what happens in the real world. The above description of the world of the twenty-first century is the background against which we have assessed the validity of the model of how a 'rational' agent behaves. According to this theory, agents gather available information about an issue, process it to arrive at the best possible decision given their fixed tastes and preferences, and do so in isolation from other agents. In other words, their preferences are not affected in any way by what other agents do. Neither do they change in any way over time.

Clearly, this is not how the world works. (Readers of a certain age might reflect on their youth to compare their tastes and behaviour with those they exhibit today.) To be fair, the theory may – may – have been a reasonable approximation to reality in the late nineteenth century, when it was first being formalised. But in general it does not fit with the world in which we live now. There are many reasons for this, but the principal one is the fact that our individual tastes and preferences are not at all fixed and independent of the preferences of others. They evolve over time. And a key factor in their evolution is that we often copy the opinions, actions and choices of others.

The word 'copy', as has been stressed, is a shorthand way of describing a range of motivations for an agent changing behaviour as a direct result of the influence of other agents. The fashion motive is one. Peer pressure is another, as is the related but subtly different concept of peer acceptance. For example the more people who are obese in your social circles, your networks, the more acceptable it is for you to be obese also. People do not decide to copy others deliberately and become obese themselves, but the social pressures and influences on them not to become obese are relaxed when other people in their networks are already obese.

Even in the 1930s, as we have seen, Keynes believed that the world was sufficiently complex, sufficiently difficult to interpret, that 'we have, as a rule, only the vaguest idea of any but the most direct consequences of our acts'. Keynes argued that the very concept of rationality needed to be redefined in such a world. Copying other agents often makes sense because they might be – it does not mean that they necessarily are – better informed than we are as individuals. In the 1950s, Simon again raised the need to reassess the definition of what constitutes rational behaviour: '[T]he task is to replace the global rationality of economic man with a kind of rational behaviour which is compatible with the access to information and computational capacities that are actually possessed by organisms, including man, in the kinds of environments in which such organisms exist.' In the complex situations which characterise much of our lives, copying is a powerful way of solving this problem, of scaling down the vast dimension of choice and its subsequent consequences into a manageable rule of behaviour. We might copy different agents for different reasons on different networks, but we copy nevertheless.

We have seen in Chapter 6 that when network effects matter, the outcomes we observe have a marked degree of inequality. Positive linking can generate success on a spectacular scale. The most popular downloads on YouTube or Flickr are often literally millions of times more popular than the huge number of videos or photos which receive hardly any hits. A few large cities are very much larger than the much greater number of modestly sized ones. At any point in time, the outcomes are unequal. But we also see persistent turnover in the rankings, so that the most popular, the biggest, do not stay in that position for ever. The time scales on which turnover takes place differ considerably between popular culture and, say, the evolution of the sizes of cities. Yet turnover is always present.

At this point, we might usefully pause for reflection: this is all very well, but look how much better off we are now than in, say, 1950. Since then we have had a great deal of state activity, of public policy interventions in both social and economic problems. And these interventions have been based on the model of economically rational agents, on the assumption that agents respond solely to incentives. Network effects and copying are entirely absent from this model. Surely we have done rather well using rational theory, especially when boosted by the addition of the late-twentieth-century insights into asymmetric information and the principle of 'market failure'. Why do we need a new, positive linking perspective on policy at all?

The stark fact is that the combination of large-scale state activity and a mechanistic intellectual approach to policy making has simply not delivered anything like the success which the founding fathers of the post-Second World War social settlement imagined would be the case. Serious economic and social problems persist.

A distinguishing feature of the social and economic history of the second half of the twentieth century is the enormous rise in the role of the state throughout the Western world. Gradually, many of the functions previously within the domain of the third or private sector have been embraced within the public sector. Roosevelt's New Deal in America in the 1930s was bitterly denounced by critics at the time as being nothing less than socialism. But the percentage share of the whole economy accounted for by the spending of the Federal government was not much more than half of what it was under Ronald Reagan. The most avowedly socialist government in the history of the UK was that of Clement Attlee from 1945 to 1951. Yet the share of the public sector in the economy as a whole under Attlee was less than it was during the government of Mrs Thatcher, renowned for her robust approach to the privatisation of state activities.

Within this framework, generations of policy makers have been raised to have a mechanistic view of the world, and a checklist mentality: all that is necessary to achieve a particular set of aims is to draw up a list of policies and simply tick them off. Such an apparently dependable, predictable and controllable environment is a comforting one in which to live.

The intellectual underpinning of the burgeoning activity of the state has been provided by mainstream economics. Paradoxically, a theoretical construct which purports to establish the efficiency of the free market has justified an enormously enhanced role for the state. It is not just the sheer size of the public sector but the range of private activities which governments now try to influence or control, either through direct regulation or through exhortations to avoid behaviour deemed inappropriate by bureaucrats, such as becoming obese or drinking more alcohol than we are told is good for us.

We have seen in Chapter 3 how the concept of 'market failure', at first sight a critique of free-market economics, has provided powerful backing to state intervention. If markets, for whatever reason, are unable to function in practice as the theory suggests they should, then regulation, taxes and/or incentives of all shapes and forms are justified. They are justified in order to make the imperfect world conform to the perfect one of economic theory. Economists, have we have seen, slip all too easily into the attitude that their core theory does not merely purport to describe how the world actually is, it is a prescription for how the world *ought* to be.

The world view of free-market economic theory is precisely one in which rational agents are able to make optimal decisions and achieve the best possible outcome in any particular set of circumstances. And so behaviour can be influenced by the appropriate set of incentives selected by the authorities. Indeed, we see a vast array of taxes, subsidies, benefits, all aimed at achieving precise, detailed outcomes. And where there are obstacles to agents making the best choice, where there is 'market failure', the clever, rational planner can intervene to ensure that the world works as the theory deems it ought.

I use the word 'planner' deliberately. In the 1940s and early 1950s, at the frontiers of high economic theory, it was demonstrated that an omniscient socialist planner, by using the price mechanism as a way of deciding how resources should be allocated, could achieve results identical to those of an idealised free-market economy, but with a more egalitarian distribution of income and wealth.

There was a great deal of practical interest in planned economies at the time. Western economies had been subject to rigid restrictions and controls during the Second World War. The planned economy of the Soviet Union not only seemed to have escaped the economic crash of the 1930s, but had been instrumental in defeating Nazi Germany, the most titanic land battles of all time taking place on the Eastern Front. Socialist planning was shown – theoretically – to be just as efficient as free enterprise, and at the same time more equitable. Of course, the discrepancy between theory and practice in the planned economies of the Soviet bloc was plain for all who chose to see. But even so, as the problems of the actually existing planned economies of the Soviet bloc became more and more apparent, the intellectual belief in the ability of clever, rational planners to control social and economic outcomes remained strong in the West. Mainstream economic theory, rooted in its particular vision of rational behaviour, provided the required intellectual framework.

We have now had over sixty years of this vision of the state. It is fundamentally different from anything that preceded it in the Western world, except during the two world wars. Yet those deep social and economic problems remain. For example in the six decades since the Second World War both the average rate of unemployment and the range within which it varies are scarcely different from those in the six decades preceding it, and, of course, in 2011 the number out of work stands at a high level in most countries. If the policy planners were supposed to achieve anything, then surely it was full employment. Joblessness on a grand scale was the scourge of the West in the early 1930s. To be fair, the maximum rates of unemployment have never again reached those of the Great Depression, but taking a longer view, averaging over decades, the rates are very similar in the pre- and post-Second World War periods. In America, the pre-war average was 7 per cent compared to just under 6 per cent post-war, in the UK the two averages are virtually identical at about 5.5 per cent, whilst in Germany the average unemployment rate since the Second World War, at just over 5 per cent, is around 1 per cent *higher* than the pre-war average.

Comparing crime rates over time is difficult, but despite sharp falls since the mid-1990s in both America and Britain, crime is everywhere much higher now than it was in 1950. Income and wealth might have increased overall, but their distribution has widened dramatically. Rational planning and clever regulation designed to cope with 'market failure' did not prevent the biggest economic recession since the 1930s from taking place in 2008–9.

*

The principal cause of the failure of what we might describe as the social democratic model to achieve its objectives is not the size of the state but the intellectual framework in which it operates. At heart, from this perspective the world is seen as a machine, admittedly a complicated one, but one which can be controlled with the right pressure on this button, just the right amount of pull on that lever. It is a world in which everything can be quantified and targets can be not only set but also achieved, thanks to the cleverness of experts.

The world is simply not like this. It is much more complex, much less controllable than 'rational' planners believe. Policy is very difficult to get right.

The main reason for this is that both Keynes's and Simon's insights into the limits to computational ability apply not just to people and companies but, equally, to what we might term the agents of the state, the public servants, the regulators, the planners. They are not specially privileged in this respect, and may very well be unable to decide, even in principle, the 'optimal' strategy to achieve any particular goal. Simon demonstrated the limits to human competence in this respect even in the humble setting of the game of chess. And most social and economic problems are considerably more complicated than chess. So, even when agents are presumed to be acting autonomously, without direct reference to the behaviour of others, when they are confronted by different information, by a different set of incentives as a result of a change in policy, their responses may be hard to anticipate. Agents may react in innovative ways, unanticipated by the planner. Or there may be entirely unintended and unforeseen adverse consequences of a change in incentives.

The problem of gauging in advance the reaction of agents to changes becomes even more difficult in situations where they base their actions, choices, opinions in part on those of others on the relevant network. So even if we know for certain how any given agent will react to a policy change now, there is no guarantee that the response will be the same tomorrow, next week, or in six months' time. The response will depend to a greater or lesser extent on how others react. This may seem obvious. But these things are not taken into account either in many ex-ante assessments of the policy terrain, or in the ex-post analysis of the impact of policies. The introduction of these fundamental features of reality into the picture rapidly leads to great uncertainty about the consequences of any given action.

At heart, policy in the West, both social and economic, rests on the framework of the economically rational agent. Agents who respond to incentives in entirely predictable ways. But unfortunately, they don't. Incentives still matter, of course, but network effects often matter even more, and in such cases they can either swamp or enhance incentive effects. We have seen, with the example of the simple market demand curve in Chapter 5 and Figure 5.2, how network effects make the evaluation of policy much more difficult than if incentives alone are operating.

Ignoring network effects when they are present can lead to seriously misleading interpretations of how things work.

Policy, whether corporate or public, needs to be looked at through the lens of the twenty-first-century model of rational agent behaviour outlined in this book. In its essentials, this is based on the principle of copying. The theory recognises that agents can innovate, make choices that have never been made before. But the guiding principle of behaviour is that agents copy the actions, opinions, choices of others according to the principle of preferential attachment. This, recall from Chapter 6, was the concept articulated by Herb Simon in his *Biometrika* article. The more an alternative is selected, the more likely it is to be selected by the next agent confronted by the choice. This, almost paradoxically, both introduces much more ex-ante uncertainty into the outcome of any given policy measure, and at the same time potentially enormously increases its potential impact through the power of positive linking.

This new model of rational behaviour is able, as we have seen, to account for two fundamental empirical features of many modern social and economic outcomes: the unequal distribution of the outcome at any point in time, and turnover within the outcome over time. It is this latter property, the ability to account in a perfectly natural way for the ubiquitous phenomenon of turnover, of popular things becoming less popular and vice versa, which is the real innovation and which previous theories are either unable to explain, or can do so only in artificial ways. In addition, it simplifies dramatically the scale of choices which agents face in the modern world. The model is an example of Simon's rules of thumb which he believed guided agent behaviour, and which he regarded as being the rational way to behave in situations of any degree of complexity.

The model of twenty-first-century rationality satisfies the basic scientific requirement of being able to explain key features of the real world. Karl Marx famously wrote, 'The philosophers have only interpreted the world, in various ways; the point is to change it.' He was completely wrong. Without a reasonably scientific way of interpreting the world, it will be very hard to change it. But it is essential that we do not fall into the trap of confusing the model *with* the world. The model of rational behaviour is merely a tool for thinking about how the world works. Like any model, it makes simplifying assumptions. It offers a way of interpreting social and economic issues which is more realistic than the economic model of rational behaviour.

*

One objection to this way of thinking is that it suggests that policy is very hard to get right. Uncertainty is a key feature of the model of twenty-first-century rationality. However, uncertainty is also a key feature of reality. Certainly, this is true ex ante, before any decision is taken. And even after the event, as the simple market demand curve in Figure 5.2 shows, network effects make evaluation of the effects difficult. This is disconcerting to many people, especially those trained in the social sciences, and in particular economics, in recent decades. The belief that clever people, with sufficient thought, really can be social engineers and design the perfect society is very deeply embedded. Real engineers really can design bridges that really work exactly as intended. The vision of society and the economy as machines encourages policy makers to take the same view of their ability to design human behaviour. But it is no longer relevant, if it ever were, to most aspects of human social and economic behaviour.

This is why the network effects view of behaviour is so challenging. I have heard frequent arguments along the lines: this is all very well, these networks may seem very clever, but they lack clear guidelines about what we should actually *do* to solve a problem. If we used the economically rational approach, we would know what to do.

The first of these points is serious and valid, and is the subject of much of the rest of this chapter. The second is an obvious non sequitur. The economic rational agent model is indeed capable of providing policy makers with an exact answer to a question: in order to achieve X, do Y. But all too often, doing Y leads to Z, or even to what we might call minus X, the complete opposite of what was intended! Recall the example of the police captain in Chapter 1. One of the most frequent and conspicuous instances of outcomes such as these is provided by financial markets. During the summer of 2011, for example, there has been constant concern about the state of Europe's economies, and the future of both the euro and the eurozone. Periodically, the French president or the German chancellor or the head of the European Commission will make a statement intended to calm the markets, or the European Central Bank will intervene in the bond market with the same intention in mind. But instead of recovering, the markets often fall further.

This way of thinking about policy does not provide control, merely the illusion of control. The twenty-first-century rational behaviour model gives a good description of how the world is. Rejecting the model as a policy tool because it may not give unequivocal recommendations is wholly invalid. The model describes how the real world actually operates, and a failure to take this into account really does mean that the success or failure of any given policy becomes a random, hit-or-miss affair.

There are other, rather more philosophical objections to a model of rational behaviour based on the principle of copying as explained by Keynes: 'Knowing that our individual judgement is worthless, we endeavour to fall back on the judgement of the rest of the world which is perhaps better informed.' At first sight, the approach seems to deny the existence of individual agency, of deliberate purpose and intent when an agent is making a decision. At a deeper level, it may even be argued that it is denying the existence of free will, the capacity of individuals to choose between right and wrong. If agents simply copy the decisions of others according to some set formula, the choice dictated by the principle of preferential attachment, then are they not merely acting as automatons?

The answer to this was implicitly provided by Keynes himself. Remember that he started his analysis from the precept that 'we have, as a rule, only the vaguest idea of any but the most direct consequences of our acts'. However, he recognised immediately that 'the necessity for action and for decision compels us as practical men to do our best to overlook this awkward fact'. Keynes argued that, to cope with situations in which we may only have a vague idea of the consequences of our acts and yet at the same time are obliged to take some decision, 'we have devised for the purpose a variety of techniques'. And the main technique, as Keynes recognised only too well, is to copy the group because the group is probably better informed than we are. We have noted a wide variety of motives which can underpin the principle of copying in addition to the one described by Keynes. But the point here, in answer to the charge that the model of behaviour based on copying denies free will and agency, is that *agents consciously choose to copy* as the basic principle of their behaviour in many situations. It is entirely rational for them to do so. They may appear to be surrendering their capacity to act as individuals, but they are deciding for themselves that this is the most sensible course to take.

A related but different criticism is that an approach to policy based upon recognising the importance of network effects will lead to companies or governments obtaining so much knowledge about individuals that they will be able to control our behaviour as never before. The terrifying vision of the world which George Orwell set out in his novel *1984*, when the citizen is powerless before the all-seeing Big Brother, is the inevitable consequence of this view of the world. Some background explanation seems necessary.

Ironically, it is the vision of the world based upon free markets and the economically rational agent to which this criticism is more pertinently applied. As we saw above, in theoretical terms, the Platonic ideas of a free-market economy and of a centrally planned economy are identical. Given sufficient information on the tastes and preferences of agents, in principle the central planner could set prices to ensure that supply and demand balanced in all markets, that the economy was in 'general equilibrium'. And so by implication, by altering prices, the planner could bring about a different equilibrium in line with his or her policy aims, or the aims of the Politburo or whatever. Agents, with their fixed tastes and preferences, can in principle be manipulated by the use of prices – incentives – to achieve any desired outcome. Of course, the contrast between the theory and the practice of the centrally planned economies was dramatic. Far from being efficient, almost to the very last days the old Soviet Union was scarcely able to feed its own population.

The revolution in communications technology does not only mean that agents are much more aware than ever before of the choices and opinions of others. It supplies companies and governments with stupendous amounts of information on how people actually behave. Some readers of this book may have bought it after being told about it by a recommender system. Such systems hold information on your previous online purchases – of books in this case – and then let you know when a book comes out which appears to fit in with your tastes. Recommender systems themselves are pretty big business. Teams of high-powered mathematicians* are competing to find better and better ways of discovering the tastes and preferences of agents based upon their previous purchases, or even previous search behaviour. The more they can discover, the better able are companies to make recommendations which are even more closely in line with an individual's tastes, and so the more likely he or she is to purchase the recommended product.

So far, all this is compatible with the view of the world encapsulated by the economically rational agent, operating in isolation from others. In such a world, recommender algorithms may indeed be able to discover over time the complete preferences of an agent, assuming these remain fixed. However, the mathematicians who devise these algorithms are also perfectly aware of the principle of copying, of the fact that a product can experience positive feedback. Once it achieves a certain level of popularity, a product, a service, an idea, may become even more popular simply because it is popular. Sophisticated strategies are developed to try to ensure that, say, a particular website comes high in the list of searches – remember that the top three sites on a Google search typically get 98 per cent of the hits. Equally, the search engines attempt to prevent their sites being gamed in this way. Behind the innocuous facade of internet searches and recommendations, teams of some of the world's best young mathematicians are pitted in a relentless struggle against each other.

Attempts to sway opinions, and possibly as a result subsequent behaviour, are by no means confined to sophisticated mathematical

---

* The maths of this is pretty hair raising. 'Quantum state diffusion', for example, is one of the concepts used in some approaches. A flavour of the difficulties can be obtained by following up some of the technical papers cited on the Wikipedia entry for 'recommender algorithms'.

manipulations. Hoteliers, restaurant owners, authors, are frequently exposed as having either posted numerous favourable reviews of
their own products on internet comparison sites, or, in some cases,
concocting highly unflattering reviews of their rivals.   Rather like
Eliza Doolittle in George Bernard Shaw's play Pygmalion, who
had been speaking grammar for years without knowing what it
was, these characters, too, in their own rather haphazard way, have
discovered the potential of positive linking without knowing anything at all about the underlying theory.

The same techniques of discovery, of monitoring individuals
and their tastes and preferences, are, of course, available to governments. The fears associated with the view that we are now in
a world where companies may know more about an individual's
preferences than the individual can articulate are magnified many
fold when the potential powers of government are considered.
And might governments in some way use the power of networks,
the tendency for positive feedback to spread ideas or behaviour
through a network, for sinister purposes of control?

Again, these issues are features of the real world. The twenty-first-century model of rational behaviour simply tells us, as a
good initial approximation, how agents behave in such a world.
It does not create the world, it describes how the world operates.
Any worries which we have are caused not by the model, but
by reality. But increased information about agents, whether in
the hands of firms or of governments, can actually be beneficial.
Alerting agents to products which they might like can be seen as
a dangerous form of manipulation, but it can also be regarded as
reducing the costs to the individual of discovering such things for
themselves.

For example a better understanding of vehicle movements
might enable traffic systems to be designed more effectively to
deliver more efficient flows. As we saw previously, Dirk Helbing

has increased dramatically our understanding of pedestrian flows, and provided insights into how to reduce the chances of people being crushed in panics. In normal circumstances, people exiting a stadium or a theatre do so in an orderly fashion, leaving adequate space between themselves and their neighbours, but in a panic, people both collide, slowing progress, and also exhibit strong flocking behaviour, copying what other people are doing. Even if there are exits spaced at regular intervals, there is still an inherent tendency for many people to attempt to escape by the same one. So, even in the hands of companies or governments, an increase in information about individuals can have beneficial effects.

*

Despite the masses of information that can be gleaned from technological advances in communications and electronic payment systems, there are formidable practical difficulties in the way of any attempt to organise the world in a way which matches the design of the policy maker. The challenges include knowing, or at least having some idea of, the structure of the network across which agents are influenced in the given context. We remember from Chapter 6 that there are different types of network in the social and economic worlds, and the kinds of approaches which might work on one may have little impact on another. So although we do not need complete knowledge of the network, we need to have a reasonable approximation to its basic structure.

It might reasonably be suggested that modern communications technology would allow policy makers to obtain a complete picture of a network: the mobile phone records of an individual, for example, can be known perfectly. However, such a network is, most emphatically, not much use in general in policy contexts. We need to know not the complete list of people with whom

an agent is in contact, but who on this network might actually influence the agent's behaviour, might induce him or her to
copy, might trigger the phenomenon of positive linking. And,
of course, the most influential people might not be the ones the
agent phones most often. It is the relevant social network which
is of interest, not the readily mapped networks of phone connections or emails. In short, a lot of important, and in particular qualitative, information concerning networks is effectively
impossible to know in many practical situations.

In addition to the structure of the social network, who copies from
whom, we need evidence on the willingness of agents to be persuaded, to copy the opinions, behaviours or choices of others. In practical
terms, these are formidable obstacles. But if we do obtain at least
some information on such matters, as we can recall from Chapter 6,
network theory itself gives us useful guides to the conduct of policy,
on which strategies are more likely to work than others.

In a scale-free network, we know that we need to identify the
well-connected individuals and to try by some means to induce
them to change their behaviours. In a random network, we know
that there is a critical value of the proportion of agents we need
to influence in order to encourage or mitigate the spread of a particular mode of behaviour or opinion across the network. This at
least gives us an idea of the scale of the effort required, and tells
us that money and time which is unlikely to generate the critical
mass is money and time wasted. In a small-world context, targeting our efforts is more difficult, but at least we know that it is the
long-range connectors, the agents with links across different parts
of the network, or who have connections into several relevant
networks, who are the most fruitful to target.

A key point here is that when network effects are present, the
most effective policies are unlikely to be generic, across-the-board
changes to incentives. Careful prior analysis and thoughtful tar-

geting become the order of the day. If we can get it right, or even approximately right, less can be more. Fewer resources used more intelligently can potentially lead to much more effective strategies. To positive linking.

Altering the structure of the network might itself also become a policy target, and one which could have powerful effects. We discussed in the previous chapter the problems of local areas where unemployment was not just high but practically endemic. At a very local level, even in poor towns, different public housing schemes, with residents from essentially identical socio-economic backgrounds, can exhibit quite different levels of worklessness. A culture can readily evolve in which an income from benefits supplemented by petty crime and casual labour becomes the social norm. In short, it is essential to take into account the fact that people live in a social context, and the particular circumstances of their various social networks can have a decisive influence on their decisions.

The most important way in which people find jobs is through personal contact. A vacancy is heard about through a friend, a family member, a neighbour. In turn, the fact that such individuals are the source of your information may send a signal to the prospective employer, especially from your informant who already works there. In an informal way, you are being recommended.

For professionals, the idea of networking – making personal contact as a key way of advancement – is second nature. But the same effect occurs at all levels of skill and qualification. Social networks are the single most important avenue for the individual to discover that a job vacancy exists. They are much more important than formal channels such as newspapers, the internet, recruitment agencies or public employment services. And from the point of view of the employer, the grapevine is less risky than recruiting from the open market, because they have additional information about the recruit.

Some public housing schemes may indeed have evolved as their social norm a network in which most adults of working age receive some form of state handout. Equally, however, the network of connections of the residents to the world of employment may just be too sparse. They simply do not hear about vacancies because not enough of them are in the loop, as it were. And we have seen in Chapter 5 that if a network has few connections, few links, then percolation across it is likely to be very limited. So policy in this instance should be directed towards increasing the social connections of the residents with the world of work, of altering the structure of the network so that it is easier for information about job vacancies to spread amongst the workless residents. And at the same time, the stronger these connections become, the greater the chance that a different social norm, that of being in work, even if it is low paid, will spread. Exactly how this is achieved, or attempted, will depend a great deal upon the purely local circumstances, of particular knowledge of the area – a theme of localisation, of devolving decisions as far as possible, which will be taken up again later in the chapter.

*

We do have the great advantage of a reliable indicator of whether network effects are present in any given situation. They leave their distinct, characteristic footprint. Namely, we observe unequal outcomes. This is a sure sign that feedbacks are operating, that factors are at work which can readily intensify the initial impact of any change in the relevant system. So we can be alerted to the presence of network effects. And we know that the type of network will influence the approach to policy which we take in any particular context. We might try to alter the network, or we might attempt to exploit its particular features once we have an approximation to its structure, its type. But how do we obtain the latter?

In many situations, simple old-fashioned survey data can do the trick, can help us approximate the type of network we are dealing with. The results need to be used in conjunction with network theory, but survey evidence can tell us a lot. This is just one illustration of how progress is being made in understanding network effects, but it is perhaps worth looking at a particular example. Social problems in particular are amenable to this way of identifying network effects and the type of network which is operating. The technique discussed in this example could be applied in a wide variety of situations.

The example I would like to focus on is a contemporary problem in British towns and cities, especially on Friday and Saturday nights: binge drinking. This involves the rapid consumption of almost unbelievable amounts of alcohol, such as a half-litre of vodka to start the evening off before getting down to more serious drinking. The results are entirely predictable. Young people – and a new aspect of this is that women are involved just as much as men – collapse incapable into the gutter, get into fights, require hospital treatment, stomachs pumped, wounds stitched, to say nothing of the inconvenience and distress this causes to respectable citizens on a night out.

Many incentive-based policies are suggested to try to deal with this problem, such as putting up the price of strong drink and restricting 'happy hours' when alcohol is served cheaply. There has always been a small percentage of young men who get into such a state, but it has become much more widespread in recent years and, as noted already, young women have become eager participants. There is no obvious correlation between the rise of this type of behaviour and factors such as the price of alcohol, which has certainly not become dramatically cheaper. So it seems very likely that a network effect is present. Binge drinking started to become that bit more popular. As a result the behaviour was copied even more, and instead of being frowned upon, it gained peer acceptance.

A survey was carried out by the marketing survey company FDS in which young people were asked to categorise themselves as binge drinkers or not. Getting people to reveal their drinking patterns is notoriously difficult, so there were plenty of cross-checks in the questionnaire, my personal favourite being the question: 'Have you ever woken up next to someone you did not recognise?' – a pretty reliable guide to alcoholic excess the night before! Nearly 20 per cent of young British adults were classified as regular binge drinkers in the survey. More interesting, however, was the way in which the participants classified their family members, work colleagues and friends. Table 9.1 shows the results of one such question: 'Are your friends binge drinkers?' We do not need statistical theory to tell us that there is a great deal of difference between the behaviour of the friends of binge drinkers and the friends of non-binge drinkers. Far more of the former themselves participate in the activity.

Proportion of friends thought to be binge drinkers	Proportion (%) for binge drinkers	Proportion (%) for non-binge drinkers
All or almost all of them	54	17
Most of them	31	24
Some of them	12	36
Hardly any or none of them	3	22

Table 9.1 Binge drinking habits of friends of binge and non-binge drinkers

The table certainly suggests the existence of network effects, of binge drinkers being influenced by the behaviour of their friends, and the non-binge drinkers being restrained by their set of friends. Of course, there are qualifications to this. We have no direct evidence on the behaviour of the binge drinkers' friends, but rely on the perceptions which the drinkers themselves have about them. There is therefore a risk that the respondents exhibit a certain amount of cognitive dissonance about their own behaviour in order to rationalise it and to protect their self-image. That said, the results do have a basic plausibility, since drinking, after all, is in general a social activity. Further, it is possible that friendship groups are formed on the basis of attitudes towards drinking. But it would be curious, to say the least, if large numbers of young people had suddenly decided quite independently of each other to binge drink, and then had happened to congregate together in friendship networks. So whilst the existence of a copying effect amongst friendship networks is not technically proved by these results, it seems a far more likely explanation than the alternative.

The results in Table 9.1 are interesting in their own right. But we can make even more use of them to obtain an approximation to the type of network structure which is operating in this particular context. Essentially, we set up mathematical models populated by agents who potentially copy each other's behaviour, and use them to see which type of network is best able to approximate the friendship structures of the two groups described in Table 9.1. We start the model off with everyone being a non-binge drinker, exactly as Duncan Watts did in his model described in Chapter 5. A few agents are chosen at random to become binge drinkers, and we follow the percolation of behaviour across the network and examine the match between the friendship patterns in Table 9.1 and what happens in the model with different networks.

I developed this methodology for a conference organised by

the US Office of Naval Research, and used data on individuals with and without bank accounts to illustrate the approach. Most people without access to financial services were receiving some form of benefit, but otherwise were very hard to distinguish in terms of their socio-economic characteristics from apparently similar people who did have bank accounts. But whether members of their friends or family did or did not have bank accounts gave a powerful explanation of whether or not any given individual would do so. Other social phenomena such as child obesity can be analysed using the same method. This description is rather cryptic, almost out of necessity, but technical details are provided in the articles referred to in the 'Further Reading' appendix.

Our analysis indicated that the network structure which best accounts for the binge-drinking phenomenon is that of the small world, or overlapping 'friends of friends', which is perhaps not surprising in this particular context. Now, the knowledge of this does not tell the policy maker exactly what to do to get a grip on this unpleasant problem, to solve it through positive linking, but it sets the parameters within which policy needs to be made. We know that we are facing a problem where network effects are important, and we have a good approximation to the type of network which is involved. If we are in some way able to penetrate the network, policy can be highly effective in reducing dramatically the scale of the problem. If we understand neither of these things, we are simply looking through a glass darkly, scarecely aware of the way in which the issue has arisen.

*

So, we have some good ways of knowing whether network effects are present in any particular situation. We have methods to approximate the network structure, the type of network which appears to be operating. We have guidelines for policy approaches once we know the nature of the network. What else do we have to help the policy maker across this difficult and challenging terrain? Again, just to repeat for emphasis, it is not the network approach to modelling which makes life problematic for policy makers, but reality itself. The network approach best describes the real world.

Importantly, network analysis is a scientific discipline which is still in its infancy. It is only in the past ten or fifteen years that the concept has really taken off and been applied to real-world systems, in both the natural and social sciences. So there is a great deal we do not yet know. We do, however, now have glimpses into two key aspects of network effects in addition to the general concepts and guidelines summarised in the previous paragraph.

The one about which least is known is the question of ex-ante prediction of percolation across a network. In other words, imagine that we are in a situation where we know that network effects are important. We believe we have a policy, perhaps a change in incentives, which might persuade a few agents to alter their behaviour or opinion; how far will this spread across the network? Will it remain confined to a relatively small number, perhaps those who are initially persuaded and a handful of agents who are very close to them, or will it cascade across the network as a whole and lead to radical alterations in the behaviour of the group of agents in the network, to positive linking?

Any attempt to give a precise answer to this question encounters a deep-seated problem. Namely, the robust yet fragile nature of networks, one of the concepts we came across in the opening chapter. We saw an application of the idea in Duncan Watts's simple but profound model, with the results plotted in Figure 5.1. The collection of individuals who make up a network will, most of the time, exhibit stability with respect to most of the 'shocks' the network receives when a few agents change their opinion or their behaviour. The system is stable in the sense that most shocks make very little difference, they are absorbed, shrugged off, and few other agents change either their minds or their behaviour as a result. So the network is 'robust'. But, every so often, a particular shock may have a dramatic effect. So the network is also 'fragile'. The behaviour of individuals across the whole, or almost the whole, of any particular network might be altered.

Of course, if we had complete information about the structure of the network, exactly who had the potential to influence whom, and exactly how persuadable each agent was, we would be able to give an exact answer to the following question: If we get a few people to change their behaviour, how far will this change spread? But then we would no longer have a model, but reality itself. It would be as if we were on a mountain, and our map was on a scale of 1:1, showing every single detail of the terrain. Obviously, even if such a map could be constructed, it would be wholly useless in practice. Imagine trying to open it to find directions struggling in a gale on top of a mist-shrouded peak! Moreover, and importantly, unlike the hills, human networks can change over time. A map that is completely accurate one year may be out of date the next.

The more information we have, the better. But in practical circumstances, what do we need to know and how much can it tell us? It seems to be the case – and this is about as much as we can say in the present state of scientific knowledge – that what really matters is how persuadable are the agents who are directly connected to the agents who initially alter their behaviour. In general, it is not so much how many agents are potentially influenced by members of this initial group, although this can be a significant factor in scale-free networks with their highly connected hubs. It is more a matter of how easy it is for the initial group to persuade the agents to whom they are connected to change their minds as well. If these are firm minded individuals, much less inclined to copy than the average person, then the change in behaviour will fizzle out, the change will not percolate across the network. But if they are keen to experiment, easily persuaded to try something different, then the behavioural change starts to spread out across the network. There is no guarantee that it will get very much beyond this second tier of agents, as it were, but it at least it has a chance.

Very recent work also suggests that the positioning in the network of the agents who initially change their minds is important. If they are either very close to each other or widely scattered, there is more chance of the initial impact spreading throughout the network. But these are rather esoteric concepts, and translating them into positive linking practice will be far from straightforward. We do know that if we can somehow acquire information on that 'second tier', the agents who are potentially influenced directly by the initial group we manage to persuade by a change in policy, we will be getting some way towards understanding how far the change is likely to spread. But even this kind of information may not be easy to obtain in practice.

We are on somewhat firmer ground in predicting the eventual outcome once an initial change has taken place in agent behaviour, as discussed in Chapter 7. Somewhat paradoxically, the very network effects which make purely ex-ante prediction difficult increase our chances of predicting how far the change will spread once we have some initial evidence on the early effects of the change. The trick is not to look for any inherent qualities or attributes of the change in which we are interested; it is rather to interrogate the data and see how strong the network effects are. This would apply equally whether we were marketing a new variety of chocolate bar or trying to get people in social housing schemes off benefit and back into work.

In a world in which network effects are important, the early stages of any evolutionary process, such as how the popularities of different choices evolve over time, are often decisive. So we can obtain a pretty good reading on where to concentrate our future resources, where to reinforce potential success in order to amplify it, once we have some initial evidence on where a change in behaviour is beginning to gain traction.

This certainly has a practical policy implication in terms of tapping into the potential of positive linking. Rather than putting all our eggs in one basket, with the misplaced confidence of the central planner, and declaring that one size fits all, we should instead experiment with lots of variants around any particular theme. Companies in fast-moving consumer goods markets such as confectionery already implement this kind of strategy intuitively. The challenge is how to go about this with public policy, a theme to which I return below. But the inescapable fact about social and economic systems in which network effects are important is that there is inherent uncertainty. Uncertainty about the future course of the economy or a social issue, and uncertainty about the impact of any policy change designed to alter the future. Even with the vast proliferation of information from the new communications technologies, and even with the advances which are being made in understanding the mathematical properties of networks, these uncertainties will remain.

An important reason was anticipated by science-fiction author Isaac Asimov in his great 1950s *Foundation* series. Asimov imagined the development of a science which combined history, sociology and maths to make general predictions about the future behaviour of large groups of people – almost exactly the challenge which now faces the social sciences when we start to incorporate network effects into our thinking. Asimov called his invention 'psychohistory', designed by the character Hari Seldon, who predicted the collapse of his civilisation, the Galactic Empire. The intricate details of the subsequent plot need not concern us, but Asimov stated that a necessary condition for Seldon's equations to work is that the population should remain in ignorance of the results of the application of psychohistorical analyses.

Here, in a nutshell, is a fundamental issue confronting any attempt to exercise precise control over the behaviour of human beings. If people become aware that attempts are being made to manipulate them, they have the capacity to alter their behaviour. And the collective imagination, the creativity of a group of humans, exceeds the ability of the would-be planner to anticipate how their behaviour might change. So, in the examples of the opening chapter, the police captain firing his gun into the air imagined that this would disperse the crowd of would-be rioters. Queen Mary thought that a policy of burning recalcitrant Protestants to death would persuade them to either flee the country or recant. Both were perfectly reasonable views to hold ex ante, but both incorrectly anticipated how people would actually react.

As we have seen, great social scientists in the past emphasised the inherent limits to our knowledge of human systems. Simon argued that the number of future paths open to us at any point in time was so vast that it made no sense at all to speak of the best, the optimal, decision. Keynes, as we have seen, stated that 'we have, as a rule, only the vaguest idea of any but the most direct consequences of our acts'. Hayek, whom we met in Chapter 6, was another great polymath social scientist who, like Simon, received the Nobel Prize in economics. Although he and Keynes clashed on specific issues of economic policy, they shared the vision of the limits to knowledge. Indeed, his Nobel lecture was entitled 'The Pretence of Knowledge'. Hayek is often seen as a free-market economist. But he was deeply critical of mainstream economics, precisely because its fundamental building block, the rational economic agent, was assumed to possess knowledge and capabilities which Hayek regarded as absurd.

A crucial point is that these limits apply to everyone. They apply to politicians, to government officials, to central bankers, to national and international regulators, to everyone. This is a very difficult point for policy makers to accept. The mentality of the central planner remains pervasive across the governments of the West, the belief that the bureaucrat is able to draw on special knowledge, special information, maybe even special gifts and talents – certainly many of them think they possess these! That, ultimately, the bureaucrat knows considerably better than the ordinary person what does and does not work.

Our current political institutions are to a large extent based on the vision of society and the economy operating like machines, populated by economically rational agents. This view of the world leads to centralised bureaucracies and centralised decision making. We live in a society where decisions are made through several layers of bureaucracy, in both the public and private sectors. On the whole, this leads to decisions that are insensitive to local (micro) conditions, and which are insensitive to society as it changes.

A lack of both resilience and robustness is a characteristic feature of such approaches to both social and economic management. Structures, rules, regulations, incentives are put in place in the belief that a desired outcome can be achieved, that a potential crisis can be predicted and forestalled by such policies. As the financial crisis from 2007 onwards illustrates only too well, this view of the world is ill suited to creating systems which are resilient when unexpected shocks occur, and which exhibit robustness in their ability to recover from the shock. The focus of policy needs to shift away from prediction and control. We can never predict the unpredictable. Instead, we need systems which exhibit resilience and robustness, which can respond well to unpredictable future events, which can recover through the strengths of positive linking.

*

Policy decisions – both in the political and corporate worlds – are largely based on thinking that emerged out of the European Enlightenment. Social sciences have been heavily influenced (understandably) by the scientific method and the tools and techniques developed in classical physics during the nineteenth and early twentieth centuries. The Enlightenment was essentially about the application of rational analysis to a sophisticated machine, in order to bring about a better world. This cognitive framework might not only have been necessary but could even have been essential to create the world in which we now live, a world of wholly unprecedented prosperity. But, as we have seen, the basic assumptions behind the social science which underpinned this vision, that of the economically rational agent, are out of date in the context of the globally integrated world of the twenty-first century.

Almost paradoxically, the global and networked society of the twenty-first century requires an emphasis on devolving decision-making power as far as is practical, a concept often referred to as 'subsidiarity'. Much tacit knowledge, knowledge derived from experience which is hard if not impossible to codify systematically, rests with individuals at a local decision-making level. Even in a large, mass-production factory, the individual workers know things about their particular task which are hard to write down. And, when telephoning a company to make a complaint, would you rather deal with a complicated rule-based system which requires you to press a sequence of digits to get a response, or would you rather speak directly to a real person, someone who might actually understand what your complaint is about? As we have seen, network effects require as much knowledge as we can possibly gather as we struggle with the uncertainties they bring and attempt to generate the benefits of positive linking.

This concept applies both within and outside any institutional structure. Take banking: most lenders apply rigid, centrally determined rules as to whether a loan can be granted, or whether an existing one is running into problems and should be called in. In contrast, Svenska Handelsbanken, Sweden's largest bank, devolves a great deal of autonomy right down to the individual officer in the local branch. In turn, these officials have an informal network across which they consult and seek advice about the creditworthiness of a company, the prospects for an industrial sector. In the early 1990s Sweden experienced a localised economic recession which was even more severe than that of the Great Depression of the 1930s. Corporate loan defaults rose sharply, but nowhere near as sharply amongst Handelsbanken borrowers as in the rest of the Swedish banking sector. The recession of 2008–9 led to exactly the same experience. One set of good results could be down to chance. Two, and it looks as if they have their devolved system and the use of informal, tacit knowledge to thank.

The devolution of decision making is also more likely to convey legitimacy on whatever decisions are made. As we have seen, agents are capable of responding in seemingly perverse ways – perverse, that is, when viewed from the perspective of economic rationality – to policy changes. Occasionally such perversity is deliberate, and intended to frustrate the intention of the policy maker. But the greater the local involvement, the greater the legitimacy is likely to be, and the more likely people will respond favourably to some policy change.

Whilst subsidiarity is usually thought of as a question of

decision-making power, there is an equivalent question about responsibility. Many believe that 'something should be done about social problem X, and since the national government is very powerful and is supposed to represent us, it should be responsible for sorting X out'. The machine metaphor approach means we have tended to think the state does have significant power – buttons, levers, the full panoply of control. This has increasingly encouraged a moral hazard-type effect, with people ceding a lot of individual and group responsibility to the state. Our understanding of networks allows us to see that the state is less powerful than we had thought. This in turn raises the question of distribution of responsibility in society. Communities ought to take more responsibility for outcomes, for their own benefit, rather than relying upon some external, magical power to solve the problem. They themselves are in a much better position to realise the benefits of positive linking that some remote, would-be bureaucrat still burdened with the mind-set of a central planner.

This is not an argument for no government, it is an argument for different government. A recurring and legitimate question posed by policy makers who have been exposed to network-type approaches is, 'Where is the silver bullet?' In other words, what is the action, the crucial decision, which will penetrate to the heart of whatever problem is being considered? The silver bullet of twenty-first-century networked reality is simple: it is to realise that there is no silver bullet. Devolution of decisions, experimentation and, yes, the recognition that many of these experiments will fail, will not deliver in terms of value for money. But experiment will discover the approach which really does work, which kicks in the network effects to deliver changes which are far more substantial than could be achieved by incentives alone, which delivers positive linking. It is a totally different approach to public policy from the one that has been in almost exclusive use throughout the West since the Second World War.

Think of some of the major challenges facing the world at present, and then think how successful the rational, central planning approach to public policy has been. Let us take as an example the management of economic crises. The rational paradigm has spectacularly failed to explain and predict not just how the present crisis emerged, but how it spread and burgeoned, initially starting with a relatively small US mortgage crisis, quickly developing into a credit and banking crisis, leading to a world-wide financial crisis, a global debt crisis for small and then big countries such as Italy, and finally, at the time of writing, into a general political crisis that is threatening the economic stability of the entire planet and the welfare of everyone on it.

Inequality is a major concern across several dimensions of the concept. The most obvious is the increase in inequalities of wealth and income in the West in recent decades. But there is also the fact that outcomes in health care from hospitals, for example, or in education from schools are 'unequal' in the key sense that they differ at any point in time. But how does inequality arise? Is some degree of inequality inevitable in a complex world of networked agents, and, if so, what can be done to mitigate it?

Inadequate supplies of food, water and energy have led to social, economic and security-related problems in many parts of the world. One crucial issue here is what economists call the 'tragedy of the commons'. If everyone has access to a particular resource how do we prevent it from being exploited to the point of exhaustion? We saw in Chapter 7 anthropologist Steve Lansing's demonstration of how Balinese rice farmers created structures and institutions at the local level to produce viable solutions. Political scientist Elinor Ostrom has showed more generally how societies evolve ways of dealing with the 'tragedy of the commons' to give outcomes which are quite different from those predicted by economic theory. Her economics Nobel Prize awarded in 2009 was the subject of many vituperative attacks from economists unable to escape the constraints of their own model of so-called rational behaviour – 'She is not even an economist!' was a cry repeated many times.

We do not yet have the answers to such shortages, or to many other major problems, but an approach to them based on an intellectual framework which recognises the paramount importance and influence of network effects is at least in tune with the real world as it exists today. It is the one which is more likely to deliver success – to discover what works in the twenty-first century and to deliver the benefits of positive linking.

Suggestions for Further Reading

This is not intended to be a comprehensive set of references covering every single point made in the text. Rather, it is a mixture, documenting in more detail some of the works mentioned in the main body of the book, and giving suggestions for interested readers to pursue. By the very nature of the subject, a substantial part of the material is not readily accessible to readers lacking some familiarity with mathematics. But the level of difficulty varies, and the fact that part of an article contains maths should not always act as a deterrent.

Contrary to the bad press it often receives, Wikipedia is in general very good in this area. The site both helps visitors discover more about a particular concept or topic, and also usually provides fairly extensive lists of further reading, of varying degrees of technicality. I have referred to Wikipedia at various points in the text, and I definitely encourage its use. Not least, the entry 'Social network' opens up a wide range of interesting material.

My own website, www.paulormerod.com, contains a considerable amount of work I have drawn on for this book. Even for academic journals, I endeavour to write in as clear and accessible way as possible, although there is often an inevitable level of technicality involved.

In terms of further web-based sources, the Cornell University Library site http://arxiv.org/ is a very extensive archive of e-prints of scientific papers. The section 'Physics and Society' lists a large number of interesting articles, although the mathematical level is in general very high and prior familiarity with the material is an advantage. The papers on the site are at the very forefront of our scientific knowledge about networks in society.

There are many books on networks, but Duncan Watts's latest, *Everything is Obvious: Once You Know the Answer* (Crown Publishing Group, 2011) is written in English and is definitely recommended. Two good sources of more technical material are Mark Newman and Duncan Watts's *The Structure and Dynamics of Networks* (Princeton University Press, 2006), which contains some classic articles, and Newman's own *Networks: An Introduction* (Oxford University Press, 2010) which is perhaps as close as you can get to a textbook in this field.

Many of the references below are to academic papers, and a judicious search of the web should enable them to be accessed for free. For example academics often post on their home pages earlier versions of the papers which are eventually accepted for publication, and so it is not always necessary to download the article exactly as published, for which many journals charge a fee. The example of the soccer hooligans in Chapter 1 is amplified in a very interesting and at the same time amusing way by Bill Buford in his book *Among the Thugs: The Experience, and the Seduction, of Crowd Violence* (W. W. Norton, 1992). There is a vast amount of material on the burnings carried out under Queen Mary. Eamon Duffy's *Fires of Faith: Catholic England Under Mary Tudor* (Yale University Press, 2009) is a well-written account by a leading historian, particularly interesting in this context because the author gives a sympathetic rationale for Mary's policy. With the historian Andrew Roach, I wrote a network-based analysis of the policy 'Emergent Scale-free Social Networks in History: Burning and the Rise of English Protestantism', published in 2008 in the web-based journal *Cultural Science* (http://culturalscience.org/journal/index.php/culturalscience/issue/view/1). We also traced how the original Inquisition gradually learned how to use networks to suppress the Cathar heresy in the thirteenth century, the first major challenge to Catholic orthodoxy in almost 1,000 years, in 'The Medieval Inquisition: Scale-free Networks and the Suppression of Heresy', *Physica A*, vol. 339 (2004), pp. 645–52. The more technical sources used in Chapter 2 are J. Adda and F. Cornaglia, 'Taxes, Cigarette Consumption and Smoking Intensity', *American Economic Review*, vol. 96 (2006), pp. 1013–28; N. A. Christakis and J. H. Fowler, 'The Spread of Obesity in a Large Social Network over 32 years', *New England Journal of Medicine*, vol. 357 (2007), pp. 370–9; and N. A. Christakis and J. H. Fowler, 'The Collective Dynamics of Smoking in a Large Social Network', *New England Journal of Medicine*, vol. 358 (2008), pp. 2249–58. The paper on crime by Steve Levitt, although it appears in a top economics journal, is very accessible to the general reader. It refers to evidence on why crime fell in America in the 1990s, although its main points remain directly relevant to crime since then. It is available in the *Journal of Economic Perspectives*, Winter 2004.

George Akerlof and Joe Stiglitz were instrumental in introducing the concept of imperfect information into modern economic theory. The seminal article is probably Akerlof's 'The Market for "Lemons": Quality Uncertainty and the Market Mechanism', *Quarterly Journal of Economics*, vol. 84 (1970), pp. 488–500, and an example of Stiglitz's work in this area is his article with Steven Salop, 'Bargains and Ripoffs: A Model of Monopolistically Competitive Price Dispersion', *Review of Economic Studies*, vol. 44 (1977), pp. 493–510.

The experimental and behavioural economists Vernon Smith and Daniel Kahneman each wrote very accessible lectures when receiving their Nobel Prizes. These were both published in vol.

93 of the *American Economic Review* (2003): Smith's is entitled 'Constructivist and Ecological Rationality in Economics' (pp. 465–508) and Kahneman's is 'Maps of Bounded Rationality: Psychology for Behavioral Economics' (pp. 1449–75). As discussed in the text, in my view possibly the most important article published in the social sciences in the second half of the twentieth century is Herbert Simon's 'A Behavioral Model of Rational Choice', in the *Quarterly Journal of Economics*, vol. 69 (1955), pp. 99–118. It is quite intricate in places, but much of the paper is in English and not maths. Friedrich Hayek does not feature as prominently in this particular book as does Simon, but his 1974 Noble lecture, 'The Pretence to Knowledge' (www.nobelprize.org/nobel_prizes/economics/laureates/1974/hayek-lecture.html), is also a seminal paper.

It is worth reading Keynes's reflections on his great book from 1936, *The General Theory of Interest, Employment and Money*, in the following year's *Quarterly Journal of Economics*, especially in the light of the economic crisis of 2007 onwards. A huge amount has been written about this, and an expanded version of Chapter 4 is in my article 'The Current Crisis and the Culpability of Macroeconomic Theory', *Journal of the British Academy of Social Sciences*, vol. 5 (2010), pp. 5–19. The other key article referred to by Simon – he wrote many more – is densely mathematical, and can be found in *Biometrika* (1955), entitled 'On a Class of Skew Distribution Functions'. This sets out a mathematical basis for the concept of copying and derives some implications. Classic psychological references to why people copy are much more accessible and include Solomon Asch, 'Opinions and Social Pressure', in *Scientific American*, coincidentally also in 1955, and S. Moscovici, E. Lage and M. Naffrechoux, 'Influences of a Consistent Minority on the Responses of a Majority in a Colour Perception Task', *Sociometry*, vol. 32 (1969), pp. 365–80. An important recent article on the effectiveness of copying is in the austere journal *Science* of 9 April 2010. Although the text is dense, it is in English and can be understood by non-specialists. As is often the case with such articles, there are many authors, but the first two named are Rendell and Boyd and it is called 'Why Copy Others? Insights from the Social Learning Strategies Tournament'. The anthropologist Steve Lansing's *Perfect Order: A Thousand Years in Bali* (Princeton University Press, 2006) offers very interesting reflections on copying and how to sustain cooperation.

Duncan Watts has done a great deal of interesting and original work on networks and copying, and the two articles discussed at some length in the text are M. J. Salganik, P. S. Dodds, and D. J. Watts, 'Experimental Study of Inequality and Unpredictability in an Artificial Cultural Market', *Science*, vol. 331, no. 5762 (2006), pp. 854–6, and D. J. Watts, 'A Simple Model of Global Cascades on Random Networks', *Proceedings of the National Academy of Science*, vol. 99, no. 9 (2002), pp. 5766–71.

Finally, a selection of my own articles on networks. One of the first I wrote which established the empirical importance of networks in people's choices was on why some people in the UK did not have bank accounts, and was written with my wife Pamela Meadows. It is 'Social Networks: Their Role in Access to Financial Services in Britain', *National Institute Economic Review*, July 2004. I formalised the procedure for approximating network structure from very limited information in 'Extracting Deep Knowledge from Limited Information', *Physica A*, vol. 378 (2007), pp. 48–52. Three more recent articles are one with Greg Wiltshire, 'Binge Drinking in the UK: A Social Network Phenomenon', *Mind and Society*, vol. 8 (2009), pp. 135–52; R. A. Bentley, P. Ormerod and M. Batty, 'Evolving Social Influence in Large Populations', *Behavioral Ecology and Sociobiology*, vol. 65 (2011), pp. 537–46; and R. A. Bentley, P. Ormerod and S. J. Shennan, 'Population-level Neutral Model Already Explains Linguistic Patterns', *Proceedings of the Royal Society B*, vol. 278, no. 1713 (2011), pp. 1770–2.

Index

72–3; criminal justice system, 50; DSGE models, 117; economic rational agent model, 269, 291; financial crisis (2008–9), 99, 101–2, 123, 291; incentives, 31–2, 37–8, 45–50; levels of uncertainties, 169, 191, 227, 285; limits to computational ability, 265–6, 287; mechanistic view of world, 262, 268; network effects, 3–4, 18, 147, 153–4, 163, 167–8, 227; network structure, 274–6, 281; network types, 168–9, 182, 274, 277; networks, 32, 35, 191, 282; 'nudge' effects, 18, 81; positive linking, 11–12, 155, 227, 267, 285, 290; price and 'bandwagon' effects, 151–3; problems, 34–5, 98, 274; understanding networks and incentives, 36–7, 98, 148–9, 266–7; Watts's model, 146–7

political: campaigns, 185, 186–7; economy, 68; ideologies, 20, 38, 97, 165; institutions, 287; policy decisions, 288

Polya urns, 224–5, 243

popularity: baby names, 255; Google searches, 164, 167, 272; music download experiment, 216–20, 222; network structure and, 132, 165; objective quality and, 130–1; preferential attachment, 166–7, 244; Simon's work, 166, 193; turnover in rankings, 235–6, 244; videos on YouTube, 129–30, 132, 163–4, 235

positive linking: background for, 137; benefits, 257, 261, 267, 289, 290, 292; cascade across network, 178, 282; commercial use of, 33, 172, 191; European crisis (2011), 125, 126; financial crisis (2008–9), 125; force, 155; Japanese economy (1990s), 110; need for, 35; non-Gaussian outcomes, 163, 243; opportunities for, 156; phenomenon, 32, 275; policy making, 12, 97, 148, 153–4, 168–9, 261, 281, 288; positive feedback, 226; potential, 267, 273, 285; potential for creating change, 59, 227; practice, 284; preferential attachment, 166; provided by networks, 154; religious effects, 250, 253; skills for, 214; social learning, 213; strategy, 276

pottery, 204, 255, 258

Prayer Book, 251

prediction(s): Asimov's 'psychohistory', 285; correctness, 112, 221; changing incentives, 37; difficulties in networked world, 222, 282, 284; policy making, 287–8; rational, 112; real GDP growth (2009), 118 (Figure 4.2); software download study, 131–2

preferential attachment: copying, 239, 243–4, 267, 270; fashion, 204; process, 166–7, 169, 204, 256; rediscovery, 166, 244; scale-free networks, 169; Simon's work, 166–7, 235, 243, 267; term, 166; turnover in rankings, 244, 248

pregnancy, 80–1

Prescott, Edward, 114

price: centrally planned economy, 263, 271; credit crunch, 108; derivatives, 103–4; discount offers, 78–9; financial crisis (2008–9), 119, 148; general equilibrium theory, 72, 104–5; incentives, 45, 47–9, 103, 215, 278; interpretations of, 149–50; labour market, 43–4; market demand curve, 42–3, 150–3 (Figure 5.2); RBC models, 116; securitisation, 106–7; supply and demand, 72, 133; 'Veblen goods', 203

Principles of Economics (Marshall), 246–7

prison, 23, 50–1, 54, 152, 252

probability theory, non-linear, 224

Proceedings of the National Academy of Science, 137

productivity, random changes in, 115

prostitutes, 46–7

Protestantism: martyrs, 5, 28–9, 165, 211, 253; Mary's policy towards, 23–4, 26–7, 250, 286; networks, 27–8, 35; religious choice, 22–4, 139; restoration, 27, 250–1; variants, 251–3

Quarterly Journal of Economics, 65, 67, 94

Quarterly Journal of Experimental Psychology, 200

railways, 194

rational agent(s): concept, 60, 111, 127,